Blockchain Politics

To Cass Chideock
Neither a peasant nor a puritan

Blockchain Politics

Ideology and the Crisis of Social Trust

Kieron O'Hara

Emeritus Fellow in Electronics and Computer Science, University of Southampton, UK

Edward Elgar PUBLISHING

Cheltenham, UK • Northampton, MA, USA

Published by

Edward Elgar Publishing Limited
The Lypiatts
15 Lansdown Road
Cheltenham
Glos GL50 2JA
UK

Edward Elgar Publishing, Inc.
William Pratt House
9 Dewey Court
Northampton
Massachusetts 01060
USA

Authorised representative in the EU for GPSR queries only: Easy Access System Europe – Mustamäe tee 50, 10621 Tallinn, Estonia, gpsr.requests@easproject.com

A catalogue record for this book is available from the British Library

Library of Congress Control Number: 2025934956

This book is available electronically in the **Elgar**online
Political Science and Public Policy subject collection
https://doi.org/10.4337/9781802207934

MIX
Paper | Supporting responsible forestry
FSC® C013604
FSC www.fsc.org

ISBN 978 1 80220 792 7 (cased)
ISBN 978 1 80220 793 4 (eBook)
ISBN 978 1 0353 6961 4 (ePub)

Printed and bound by CPI Group (UK) Ltd, Croydon, CR0 4YY

Contents

Acknowledgments

Thanks to Taylor & Francis publishers (https://www.tandfonline.com/) for permission to adapt text from "Burkean conservatism, legibility and populism," *Journal of Political Ideologies*, 26(1), 2021, 81–100, https://doi.org/10.1080/13569317.2020.1844371, in Chapters 6 and 10. I have also adapted text from "Digital modernity," *Foundations and Trends in Web Science*, 9(1–2), 2022, 1–260, http://dx.doi.org/10.1561/1800000031, published by Now Publishers (https://www.nowpublishers.com/), © Kieron O'Hara, in various places, including Chapters 1 and 9.

This book was written prior to the US Presidential Election of 2024. A small number of minor changes were made in proof to recognise Donald Trump's victory (including the addition of this paragraph), but introducing no hindsight about the nature of his second administration.

My account of trust was developed independently from that of Professor Katherine Hawley, but shares many of its features; once I began work on this book, I engaged with her research and learned much from it. I hoped to contact Professor Hawley when I had finished my work to discuss some of its themes and its relationship with her own work. Sadly, she was already seriously ill when I began writing, and died tragically young in 2021.

Introduction: the importance of trust for democratic politics, and why the populist right and post-Marxist left want to eliminate it

21st century politics are utterly bizarre. Despite relatively few people engaging directly with extremes, we seem to have been transported to the 17th century world of Matthew Hopkins, Witchfinder General. On one side we have a puritan, compassion-free, vengeful, punitive regiment of unsmiling moral police determined to suppress all trace of old sinful ways and to purge wicked traditions, contemptuous of due process and untarnished by presumption of innocence. They face a pugnacious mob of snarling, swivel-eyed, snaggle-toothed peasants with pitchforks, impervious to reason, disdainful of calls for calm, scornful of authority, oppugnant to the Riot Act, and utterly convinced by their preachers that they are beset by every kind of sprite, gargoyle and demon that ever featured in a story book.

To talk of political trust in these circumstances is heroic. Like the humane, sherry-tippling squire of 1647, we are trapped in a pincer by two terrible Manichaean movements, who agree only on the immanence of the apocalyptic struggle of Good versus Evil, cooperate only to drown moderate voices in a torrent of manufactured invective. The impediment under which we labour today, unlike the hapless squire of yore, is that ideas spread far more quickly and effectively with networked digital technology. And the technology that both sides use to automate their mobs is increasingly defended as a means by which society can be run devoid of trust. It wasn't so long ago that trust was seen as the glue that held societies together and allowed them to prosper (Fukuyama 1995). Now, Puritans and Peasants alike see it as a dangerous vulnerability to be transcended.

THE PURITANS AND THE PEASANTS

The post-war period was marked by the massive growth of the intrusive state. This made the practical project of steering the large and unwieldy governmental infrastructure increasingly salient (Deutsch 1963, xxvii), as it took unprecedented shares of national income and the private property of its citizens with an efficiency and bureaucratic indifference that would have been the envy of medieval despots (McCloskey 2010, 331–332). The struggle to control the new Leviathan cultivated "a time of brittle cynicism about the activities of politicians," as it says on the back cover of Bernard Crick's brilliant *In Defence of Politics*, which advocated politics as a social activity for allocating resources and decision-making, essential to sustain a community grown too complex and plural to be governed by tradition or ideology (Crick 1964, 24).

Since its publication, the state has grown further while the political situation has evolved, becoming both more complex and less pragmatic than Crick recommended. First, there was a shift from class- and solidarity-based ideologies, to those, such as neoliberalism and Rawlsian liberalism, which valorised individuals and their choices. This, to some commentators, transformed social politics into the politics of personal prosperity, in which politicians "are exposed to pressures from a society in which the ties of habit have been loosened and agreement on what is appropriate conduct dissolved. The censure of public opinion remains, but it has become more capricious and is formed all too often by disappointment over the non-fulfilment of material expectations which it may not be within the power of the politician to satisfy" (Johnson 1980, 136; and also Urbinati 2015).

Secondly, particularly following the political protests of 1968 (Bourg 2017), and the American civil rights movement, perceptions of the purpose of politics seemed to change from the allocation of scarce resources, where the aim is to find compromise or the win-win, to rights and wrongs, where the only acceptable result is victory, and defeat is but a prelude to resistance. We might contrast the two attitudes using the terms *getting-things-done politics* and *beggar-thy-neighbour politics*. In the latter, "narratives of distrust, conspiracy, betrayal, and fear thrive in discursive vacuums … because narrative is one way of focusing and containing perceived risks and threats" (Warren 1999, 341).[1]

Thirdly, in Crick's day, the educated and the wealthy alike skewed to the right of politics, while the poorer and less-educated tended left. By the 2020s, the educated were firmly on the left (Piketty 2021), creating new cleavages between the educated elite and the unlettered masses (now bedfellows), spawning tensions within both right and left. On the left, *bien-pensant* intellectuals marginalised working class institutions and attitudes. The right

[1] This book does not consider political violence, but it seems plausible that the heightened rhetoric of beggar-thy-neighbour politics is more likely to lead to a situation where violence would appear to some political actors to be the only way of preventing outcomes branded as unacceptable.

struggled to square cultural conservatism with the free-and-easy manners of the hyper-wealthy (a tension memorably depicted in Alan Hollinghurst's novel *The Line of Beauty*).

And fourthly, especially following the collapse of the Soviet Union, it became clearer that the working class had unaccountably rejected its historical destiny as the nemesis of capitalism. This was a problem for the left: the metaphor of struggle had worked well, evoking images of two mighty forces, the bourgeoisie and proletariat, confronting each other, gaining victories, rallying after defeats, waving flags and so on. Yet this now seemed absurd, as the workers rather preferred liberty, patriotism and opportunity within a civil society.

The Puritans

A new style of leftism was needed, which gradually emerged as the intellectual elite insulated itself from the disappointing proletariat, which it now despised. It put down its Marx, and picked up its Gramsci, nursing its wounds in various cultural institutions, universities, art galleries, civil service bureaucracies and NGOs (Embery 2021, 18–22, 47–49, 98–99; Mounk 2023, 97–111; Rufo 2023, 36–52). It created and dominated an ecosystem of "pressure groups," "creatives" and "community groups" that purported to represent "civil society," so that not only did it manage to take over many branches of government administration, it also became a spurious "interface" with the citizenry (Walden 2020, 157–176).[2] For some non-obvious reason, many new leftists also became stand-up comics.

The bourgeoisie was not smashable, but it was nice enough to be embarrassed and shamed. Victimhood became the means; it was enough to show that one had been badly treated, neglected, traduced or otherwise held back in one's own opinion. Civil society was asserted to be a racket that denied social justice and diversity, and so civility was rejected in manners as well as social philosophy. In countless committees, well-meaning moderates were shouted down, and being polite and a little cowardly, they quailed. Whereas the class

---

[2] See https://dnco.com/work/transport-for-london for a priceless example, where "public consultations" about the new names of London Overground lines took the form of intensive workshops with "writers and creatives" exploring "themes of decolonisation, queer histories, intersectionality and young London's perspectives," "interviewing leading historians, academics and transport specialists delving into topics such as LGBTQ+ histories, East End migration and the fascinating world of London slang and linguistics," leavened only by "150 customer intercepts, hearing from everyday Londoners about what makes the Overground special." In other words, the "consultation" was designed to produce a desired outcome, which it duly did.

struggle had been about the competition for political and economic power, the new Puritans demanded obeisance in behaviour, language and thought, shifting from class war to culture war, and making far more progress.

Their struggle was grounded by uncompromising loathing for their own societies. Incredibly, at the beginning of the COVID pandemic, one British academic was desperate that her own university *didn't* develop a vaccine. "If my university is the first to develop the vaccine, I'm worried that it will be used as it has been in the past, to fulfil its political, patriotic function as proof of British excellence" (Cousens 2020). In the event, the excellence of British (and German, Swedish and American) pharmaceutical companies (and their multinational researchers) did result in vaccines within months, which was certainly a cause of pride, and for everyone else of celebration. Pride in the national flag might get a rock band into trouble. "Oasis are the band of choice for flag-shaggers and Reform voters – it's remarkable how often their fans have the butcher's apron [i.e. the British Union flag] on their Twitter bios, just as Noel [Gallagher] had it painted on his guitar" (Price 2024). Flag-aversion and contempt for ordinary fans, if not the crudity, are also evident in this diatribe from a graduate of Germany's "inherently racist education system."[3]

> Germany hosted the football World Cup [in 2006] and it quickly seemed to infect almost everyone around me with an enthusiasm for the alleged greatness of the reunified country. … Never before had I seen so many black-red-gold flags waving from windows, hanging in cars, painted on cheeks. All the symbolism and pride in being German that had been reserved primarily for neo-Nazis … had suddenly become mainstream. … When Germany won the World Cup in 2014, the national team were welcomed with a public victory celebration in the heart of Berlin, sponsored by major German brands and broadcast live by state channels. A journalist colleague rightly criticised this "warrior-like self-aggrandisement" of the national team … . The presence of players from dual-heritage families on the German football team doesn't really change the racist dynamic attached to this national fan culture. … I, for one, will be doing [in the 2024 Euros] what I have done since 2006: hoping that the German football team – again one of the favourites – lose their opening matches and get kicked out of the tournament as quickly as possible. (Aydemir 2024)

It is worth emphasising that the author hadn't been attacked or suffered racist abuse from football fans; she simply objected to people enthusiastically supporting their country's team.[4] Thus the Puritans.

---

[3] The author is German-born, of German-born parents, of Kurdish-Turkish extraction (https://en.wikipedia.org/wiki/Fatma_Aydemir).

[4] In the sequel, she didn't get her wish, and the Germans made it through to the quarter-finals before elimination. In praising his team, the German manager was quoted as saying "We are a country that is too sad in too many situations, with a

The Peasants

Modern society is fraught with dangers imagined and real (poverty, warfare, totalitarianism, terrorism), but the constitutional protections of liberal democracies are increasingly impenetrable. "If any power is to be effectively given to the people-as-population of a large and complex modern society, this can be done only by means of institutions and procedures that are intricate to the point of being baffling" (Canovan 2004, 245). The Peasants, or, as the German word has it, *Mistgabelmob* (pitchfork mob), arise where complexity seems to undermine common sense and straightforward moral nostrums, as when organised criminals use the painstaking safeguards of the legal system to avoid conviction, or the welfare state seems to reward those who make no contribution.

The Peasants are populists, therefore, but of a certain type. They revel in vulgarity, associating effete metropolitan ways, good taste, high culture and urbanity with the subversion or exploitation of community and communal norms. They are outraged by those who take social determinism more seriously than moral responsibility. They suspect the Brahmin class, not unreasonably, of looking down upon them, while, by virtue of its wealth and social standing, being insulated from the changes it imposes upon everyone else. Yet Peasant leaders often have impeccably Brahmin CVs: in the US, Ted Cruz went to Princeton and Harvard Law School; Steve Bannon to Georgetown University, Harvard Law School and Goldman Sachs; Ron DeSantis to Yale and Harvard Law School; and J.D. Vance to Yale Law School. For a (dated) critique of the British elite, see (Walden 2020, 72–89).

Peasant populism was facilitated by the same political changes that created the Puritans. Donald Trump may seem like he just fell out of a coconut tree, but Michael Sandel predicted the appeal of his noxious politics as early as 1996 (Sandel 1996; also Gray 1995). He is really the (hopefully) end point of a process of the decline of the Republican Party from business- and liberty-friendly establishment party to paranoid rabble, through McCarthy (Hofstadter 2008b), Goldwater (Hofstadter 2008c), Buchanan (Stanley 2012), Gingrich's Contract With America and the Tea Party.

There are different ways of defining populism, but whether we do it ideologically (pitting a homogeneous "people" against an arrogant elite – Müller 2017; Rummens 2017), strategically (in terms of the methods of gaining power, especially by a charismatic figurehead with an almost mystical connection to the "people" – Weyland 2001), or sociologically (in terms of the rhetoric and

dark perspective on things, and I hope this symbiosis can create something. We are stronger with unity, when we think about what we can do together; together with your neighbour you are stronger. This unity here felt good." The general mood was lighter than our misery-guts hoped (Connolly 2024).

practice of low politics – Ostiguy & Roberts 2016), today's Peasants get their va-va-voom from the manifest unpalatability of the Puritans. Yet the attack on "elites" in the name of the "people" undermines getting-things-done politics. Peasant populists turn out to be neither very disciplined nor good at running things, and they are most comfortable with opposition and ideological animus.

Mark Sedgwick suggested a helpful characterisation of the intellectual roots of the "radical right" (Sedgwick 2019, xxiii–xxiv), which I have adapted to apply to Peasant politics (even if we doubt that President Trump has read Spengler, Schmitt or Gramsci).

1. *Declinism*: the Spenglerian doctrine that the West is in decline.
2. *Emergency*: in the current sociopolitical context normal politics won't cut it. It is a state of emergency and we need a politics of the exceptional (an idea of Carl Schmitt).
3. *Manichaeanism*: again following Schmitt, a fundamental political division between friend and enemy.
4. *Anti-elitism*: political elites are thwarting the people.
5. *Gramscian metapolitics*: aiming for cultural hegemony (Bates 1975), establishing the Peasant view as common sense.

Thus the Peasant.

Interestingly, the Puritans can be sketched analogously, disclosing their similar roots.

1. *Rejectionism*: the West is uniquely evil, and its past actions are responsible for most of the political and geopolitical ills of our age. Alternatively, "the West" is merely a cultural construct. It has no independent existence, because "the modern West owes much of its cultural DNA to a wide range of non-European and non-white forebears" (Mac Sweeney 2023, 6; and also Quinn 2024). However, "as a conceptual framework, it has provided a justification for Western expansion and imperialism, as well as ongoing systems of White racial dominance" (Mac Sweeney 2023, 6), thereby biting the hand that fed it.
2. *Emergency*: identically to the Peasants, as in the myriad declarations of "climate emergency" (Hulme 2019).
3. *Manichaeanism*: again identically.
4. *Conflict between elites*: the present political elite dominates and oppresses minorities, and is to be replaced by a vanguard which will dismantle the *Ancien Régime*'s racist, sexist, homophobic, ableist and colonialist structures.
5. *Gramscian metapolitics*: as the Peasants, *mutatis mutandis*.

A Vital Caveat

This book is about actors on the political scene, those actively planning and seeking political effects: politicians, special advisors and think-tankers, journalists, activists (and slacktivists), social media influencers, campaigning bishops, trade unionists, demonstrators, leaders of NGOs, politicised artists and novelists, "community leaders," party members, lobbyists, politically-conscious educators and social scientists, party funders, and so on. It is *not* about voters or citizens.

You might vote for the Peasants because you are appalled by the Puritans, and do not trust mainstream politicians to keep them out; or for the Puritans as the most implacable opponents of the Peasants. You might vote Puritan because their ideas chime with your disgust for racism, homophobia and sexism, or Peasant because their messages about the failures of government and taking back control resonate. That does not make you a political actor. Voters and citizens react to debate, but they are not the target of my argument. They are not passive, and may hold some responsibility for the decline in political trust and civility, but gauging its extent is not the purpose of this book.

The characterisations are perhaps vague; Peasant and Puritan are more styles of politics, almost psychological profiles, than thought-out ideologies. It is a sensemaking exercise on my part. The main point is: neither trusts.

THE POST COLD WAR REDISCOVERY OF TRUST

The demise of the Soviet Union was an indirect spur to the discovery of trust as a sociopolitical issue. The collapse of the governing ideology in Eastern Europe revealed the lack of civil society institutions that could mediate between persons and build social trust (Krastev & Holmes 2019). Meanwhile, there were a few hubristic claims about the ideological triumph of secular liberal democratic capitalism, most famously in Francis Fukuyama's blockbuster *The End of History* (1992). This was a somewhat misunderstood book which argued that liberal democracy provided people with recognition, dignity, liberty, security, equitably-spread prosperity and public services, a set of benefits that no other form of political organisation had ever achieved or could achieve. Deirdre McCloskey made a similar point from a longer perspective when she argued that, while some democracies are thriving and others in decline, life in the West is basically pretty good (McCloskey 2010, 118–120).

Yet these advantages ended neither history nor politics. Concern shifted to what has been called social (sometimes human) capital, the basic "lubricant" of civil society and the economy (Arrow 1974, 23). Now that liberal democracy was, for the moment, the only game in town, questions emerged about how a plural, free, open and increasingly non-religious society could function without collapsing into conflict. Trust was seen to be the key, and Fukuyama's follow-up to his misfiring blockbuster was, interestingly, a discussion of social capital (1995).

The turn of the century saw a slew of worries about individualism undermining community (Putnam 2000). Fukuyama urged that "trust arises when a community shares a set of moral values in such a way as to create expectations of regular and honest behavior" (1995, 153). There was much discussion of a "crisis of trust" (Seligman 1997, 165; O'Neill 2002, viii, 4–5, 16–17; Uslaner 2002; Rotenberg 2018, 1–3; Skelton 2019, 85; Timothy 2020, 152), and some suggested that the task of political science was to understand why withdrawing trust was "a sensible response to the world" (Hardin 1999, 40). My own investigation around that time identified a few tricky areas (typically involving authority or expertise) where trust was being asked to do a lot of work, because expectations of beneficial outcomes were too high, and cynicism was an easy resort. However, I concluded then as now that across society trust was not in short supply (O'Hara 2004).

The interplay between Fukuyama's two books is interesting. The social capital that he lauded in *Trust* presupposes the order created by civil society in liberal democracies (much of his later work focused on social order). Yet intellectual and political elites scorn that foundation, blithely assuming that democracies can continue to be pleasant, tolerable and trusting, even as Puritans and Peasants, in their different ways, work to undermine the ordered, bourgeois life which furnishes our prosperity and comfort. The "end of history" was rather the prelude to a new phase of history by other means.

THE ARGUMENT

In this book, I shall look at trust as a factor in the prosecution of politics, in both its getting-things-done and beggar-thy-neighbour guises. I make no pretence of objectivity – I prefer the former style to the latter. I shall argue that many of our self-styled "progressive" ideologies are destructive of trust and aim to make it redundant. Meanwhile, a milder small-c conservatism nurtures the trust essential for getting-things-done.

Peasant-on-Puritan knockabout, with right-wing populists ranged against "progressives," has reduced the space for trust. In broad terms, there is widespread resistance to Puritan politics, with its dreams of social control and contortions of moral dilemmas and attitudes. In return, the Puritans don't trust citizens to make the right decisions about how to run their associations, manage climate change, and so on. Having colonised many institutions, particularly educational and governmental ones, they have put themselves in a position to make technocratic decisions behind the scenes.

Populism can be tough to get off the ground, because "the people" are naturally highly heterogeneous. As Hermann Hesse put it a century ago, "Who is the people? Is it a street speaker or is it those who listen to him; is it those who agree with him or those who brandish their cudgels and shout him down?" (Hesse 1972, 84). Yet a potent oppositional group can coalesce around dislike of technocracy and being bossed around, especially when there appears to be no other route to object. Because the Peasants respect no institutions, they look well-placed to confront and defy the Puritans in their institutional fortresses.

More moderate politicians who want to make institutions work to practical benefit lose appeal because the institutions, for the discontents, are the problem. Moreover, plenty of establishment moderates are Puritan fellow travellers, broadly in agreement on specific matters (such as equality, racism or climate change). They deal asymmetrically with the extremes; they build a *cordon sanitaire* around Peasant parties of the right, but, in their struggle to impose reasonable measures to discourage discrimination, they license the illiberal pursuit of excessive Puritan absolutism. Put another way, *they cannot be trusted to curb the Puritans.* The Puritans and their fellow travellers, in short, enable the Peasants, creating the conditions in which they flourish, and *vice versa.*

WHAT TRUST BUYS: THE EXAMPLE OF BREXIT

Where trust has broken down, politics can't properly be done, just as wars cannot be won if no-one ever surrenders. Brexit is one of the totemic issues dividing Puritans and Peasants, and the botch that was made of it was entirely down to the lack of trust and trustworthiness in polarised British politics.[5] The getting-things-done political task was to craft a post-EU project that would build on the advantages of returned sovereignty, while retaining good relations with the EU, the UK's nearest neighbour, largest trade partner, and closest geopolitical ally.

Why was it so bungled? I would say the lack of trust: between Leavers and the Remainer establishment; between Remainers and voters; between the Conservative Party leadership and its troops; within the Tory leadership itself; between the Tories and the Labour Party; between the Labour leadership and Parliamentary Party; between the Parliamentary Labour Party and

[5] I should declare my own position. I voted Remain, more because of the risks of leaving, than from any great love of the EU. I also opposed a referendum, an utterly flawed way of making decisions and establishing consensus (O'Hara 2006). But, as one was held and the result pretty clear, the decision to leave had to be respected, if politics was to be seen as a trustworthy profession. A second referendum should have been out of the question.

the membership; within the Remain group. For the Parliamentary mess, see Wager (2021). The EU didn't help; its sole aim was to punish Britain for leaving, thereby making a mockery of Habermas' argument (2012, 40–41; 2015, 40–42) that the Lisbon Treaty democratised the EU by making it possible to leave.

Article 50 was triggered too early by Remainer Theresa May because she felt she had to demonstrate her trustworthiness to the Leavers in her party. She tasked figures she did not trust (David Davis, Liam Fox and – demonstrably the least trustworthy politician around – Boris Johnson) with carrying out the negotiations. Davis and Johnson reneged on that trust by failing to negotiate seriously and resigning the moment a withdrawal agreement looked on the cards. May demonstrated her lack of trust in the three by sidelining them with officials. The cross-party Parliamentary Remain majority, egged on by the wretched Speaker John Bercow, got absolutely everything wrong because it was unable to unite in a purposeful coalition, making Johnson the only plausible Prime Minister by the Summer of 2019. This was a triumph of beggar-thy-neighbour politics and, in Euro-lingo, *un petit-dejeuner d'un chien.*

What might have happened, had trust not been on short rations? The political class might have pulled together, not in agreement, but behind a democratic and constitutional process. Those parties with large Leaver factions should have committed to leaving. The next election was due in 2020, giving each party time to work out a precise negotiating position to put before the electorate in its manifesto. Labour might have wanted to keep close to the EU for its social benefits, while the Tories might have promoted free trade, and UKIP could aim to cut immigration, for instance. Consistent Remainer parties like the Liberal Democrats or Scottish Nationalists might have decided to campaign for a second referendum or even to reverse the decision. It would have been a struggle in most parties to produce such manifestos, but groups of people who trust each other can make the necessary compromises.

Election called, manifestos presented, a new Parliament would have a mandate to trigger Article 50 and negotiate something specific (or to call the whole thing off). The meaning of Brexit, unclear during the referendum, would have been decided by the election, and the project would be the major task of that Parliament. The EU, wanting to punish Britain, would doubtless still have negotiated in bad faith, but the UK would at least have had clear goals and a mandate to push for them.

There are many other such areas. The response to COVID is an obvious one, which involved a complex trade-off between citizen trust of government, citizen trust of science, government trust of science and scientists, scientists' trust of government, everyone's trust of the media, as well as the Puritan/Peasant divide which in some countries even politicised the pandemic suppression measures, and on top of that the influence of social media and misinformation, and on top of that issues about the reliability of statistics in some countries, and on top of that the point that levels of uncertainty for policymakers were extraordinarily high. As this is a work of philosophy, not political science or sociology, I won't delve into this; see Devine et al (2021) for an early survey, and Seldon & Newell (2023, 181–240), Agar (2024, 345–350), Egerton (2024, 55–59) and Sylvester (2024, 152–154) for early assessments of the British response. A trusting political culture would at least allow coordinated and depoliticised responses (although even then they may be based on erroneous science).

THE STRUCTURE OF THIS BOOK

The Puritans and the Peasants are both corrosive to democratic politics. In the literature, the Peasants get the worst of it (Mounk 2023, 17–18, for quantitative evidence). Few writers of any quality are prepared to go on the record and defend Trump, Farage, Le Pen, Cruz, DeSantis, Wilders et al., while the conspiracy theories that are catnip to this bizarre tribe are easily dismissed. Academics are appalled when young people are "radicalised" by the far right, and write hand-wringing articles on how to stop it, while doing their best to radicalise their students on the Puritan side. Yet when we ask why the Peasants thrive in public life, few commentators are able to admit the unpalatable truth that they are widely preferred to the Puritans.

Hence this book is focused on the Puritans and their fellow-travellers. That does not mean that I don't condemn the Peasants; I do, but that is less of a lonely task. I have tried elsewhere to show how the small-c conservatism I outline in this book is opposed to right-wing populism (O'Hara 2021), but in general I assume without proof that the Peasants are unfit for office and inimical to trust, and therefore cannot provide solutions to the serious problems democracies face. They have no jointly coherent positive agenda, disagreeing, for example, on membership of the EU, support for Ukraine, and women's and gay rights. They also don't really know what to do about, say, immigration; they oppose it, but are unable to work out how to stop it. That makes them far less effective politically than the Puritans, who get things done thanks to their strategic "long march" through our cultural institutions (only in Eastern Europe have the Peasants successfully taken over the machinery of the state). The Peasants worry me far less than the Puritans. Hence the substantive arguments about unfitness for office and injuriousness to trust will be targeted at the Puritans only, in Part III.

It is not a binary, and I suspect relatively few people are firmly committed to either side. I tend to agree with Yascha Mounk that most citizens, ambivalent about politics, broadly want our diverse societies to succeed (2022, 18), but don't see why guilt, sanctimony and intolerance are required for the purpose.

The book falls into three parts. An opening chapter will look at political trust, and how the Puritans and the Peasants propose to wish it away. Part I of the book will review the nature of trust. This is foundational work, and could be skipped if you want to get straight into the politics, although my analysis of trust is a crucial premise for many of the political arguments. Part II will lay out a traditional Burkean small-c conservative ideology, to show how it both depends on, and fosters, trust in a tight circle. Part III will look more critically at how various prominent strands of Puritan thought undermine trust. In conclusion, the need for trust in plural and competitive democracies will be discussed, looking in particular at the political projects it would enable.

This book is itself part of a loose and unintended trilogy, the fruit of a project that grew out of hand. This is the middle part, appearing first. A conceptual opening would be an analysis of trust and how it operates in society; this is condensed into Part I here, with many lacunae, caveats and promissory notes which I hope will be made good later. The closing work would focus on the "trust dividend": what could and should be done with trust restored and shored up in Western democracies.

1. Trustless trust and the blockchain metaphor

The traditional view of representative democracy depends on trust in the representative's political expertise. In a famous speech of 1774, Edmund Burke outlined this view at a time when Britain's democracy was pretty undeveloped (also Montesquieu 1989, 11–12). He had just been elected as one of the two Whig MPs for Bristol, and his new colleague had committed to acting according to his constituents' commands, to Burke's horror.

> Certainly, Gentlemen, it ought to be the happiness and glory of a Representative, to live in the strictest union, the closest correspondence, and the most unreserved communication with his constituents. Their wishes ought to have great weight with him; their opinion high respect; their business unremitted attention. It is his duty to sacrifice his repose, his pleasures, his satisfactions, to theirs; and, above all, ever, and in all cases, to prefer their interest to his own. But, his unbiassed opinion, his mature judgement, his enlightened conscience, he ought not to sacrifice to you; to any man, or to any sett of men living. ... Your representative owes you, not his industry only, but his judgement; and he betrays, instead of serving you, if he sacrifices it to your opinion. (Burke 1999b, 10–11)

Constituents, on this view, are like beneficiaries of a trust of which the representative is a trustee (Pitkin 1967; Elster 1998). It is a rejection of beggar-thy-neighbour politics. "If Government were a matter of Will on any side, yours, without question, ought to be superior. But Government and Legislation are matters of reason and judgement, and not of inclination" (Burke 1999b, 11). The representative is to assess her constituents' interests dispassionately. She needs to engage with all strata of local society, not just her own supporters: factory owners and trade union representatives; teachers and doctors; parents and patients; environmental and protest groups; businesspeople large and small; bankers and charity workers; taxpayers and welfare recipients; read the local newspapers and monitor social media (although not the abuse). And she must learn about the community from these conversations. Complex local matters must be balanced against the national interest, and the job of the MP is both to articulate local needs within a national context, and also to explain within the constituency how wider issues impact on the local community, "provoking citizens to reflect on future perspectives and conflicts in the process of devising national politics" (Urbinati & Warren 2008, 391).

The representative performs a valuable task within democracy: fluid engagement with constituents, the party and ministers mean that "representative electoral institutions … seem a fairly good way by which a government can find out what people will do and what they will stand for" (Crick 1964, 27). Representative politics comprises "many small steps by which information is processed – steps which at first articulate interests, sound out possible points of consensus, push people into positions, test general programmatic proposals, and then bring about, for the time being, a consolidation of what is binding, by decisions about legislation, the budget, or general directions of policy" (Luhmann 2017, 60).

The delegate, on the other hand, must obey the instructions of her constituents, and as Burke had warned, "if we do not permit our members to act upon a *very* enlarged view of things; we shall at length infallibly degrade our national representation into a confused and scuffling bustle of local agency" (Burke 1881, 121, his emphasis). Today's Parliamentarians look more like delegates than Burkean representatives. "Over the last quarter century there has been growing concern among academics and political commentators that [British] MPs have become increasingly parochial in their focus, prioritising the constituency and performing a local grievance-chasing role rather than focusing on national and international issues in their role as scrutineer and legislator" (Hansard Society 2017, 31). Worse, a delegate risks taking the majority's steer, without properly representing minority interests.

TRUST AND ITS DISCONTENTS

John Locke drew a distinction between legitimate and illegitimate power. A legitimate ruler rules for the benefit of the ruled, whereas an illegitimate ruler recognises no obligations, and hence, however wise or benign, does not deserve trust because she could at any point arbitrarily change policy (Dunn 1984, 295). *Trust must engage the expectations of the trustor.*

For instance, Burke (unsuccessfully) urged the British government to stop exploiting its American colonies for tax purposes, but even had he been successful, could the colonists have trusted it? Having disregarded American opinion once, even if it moderated its policy it might capriciously change its mind again for all they knew (Crick 1964, 122). More recently, American allies feel abandoned by the isolationist "America First" doctrine, and while it flourishes, no internationalist fence-mending can succeed fully, because another isolationist might take office in the future, unrestrained by today's olive branches.

On Locke's contractual philosophy, constituents trust their representative to articulate an aggregated and balanced view of their interests, while retaining "the right and duty to judge for themselves how far their trust has been deserved and where and when it has been betrayed. And if they do judge it to have been betrayed, they have every right to act in concert and seek to re-establish for themselves a form of sovereign power in which they can, once again, rationally place their trust" (Dunn 1984, 297).

This statement is central to the issues to be discussed in this book. How far does "acting in concert" go, and what kinds of re-establishment are consistent with getting-things-done politics? It would not be acceptable to Burke if "acting in concert" included the possibility of revolutionary or unconstitutional change, but he would agree that it includes voting out and replacing an MP (as the voters of Bristol would have replaced Burke in 1780, had he stood again). But modern ideological trends are supportive of more revolutionary implications, as we shall see in Part III. For now, let us note that an expansive Lockean view invites beggar-thy-neighbour politics, turning politics into a constant effort to persuade citizens that they have been betrayed and to establish new forms of sovereign power. On the other hand, even moderate Lockeanism threatens getting-things-done politics. Policies thought necessary by politicians but unpleasant by voters may result in consequences for the former – a dilemma famously expressed by Jean-Claude Juncker: "We all know what to do, but we don't know how to get re-elected once we have done it."

One way of putting the problem is that, when the representative has power and authority, she cannot be trusted unless there is restraint on her behaviour, but her power may hinder restraint. The view that reliable trust requires effective constraints is not new. In premodern Europe, it was widely thought that trust was only reliable when underwritten by God, as untrustworthy folk would face punishment in the afterlife. "Covenants," said Thomas Hobbes, "without the Sword, are but Words, and of no strength to secure a man at all" (1968, 223). Even Locke argued in his *Letter on Toleration* that atheists had no right to toleration because they expected no supernatural retribution, and therefore might do anything (Dunn 1984, 287–290).

Talking Down Trust: the Peasants and the Puritans

Most democratic theorists respect the ability of institutions to produce legitimacy, equality, responsiveness and (usually) decent outcomes, earning the trust of voters and candidates taking part in democratic processes voluntarily, even if they don't expect to win (Lenard 2012, 38–53), and conversely that trust in institutions is also essential for democracy to function (Lenard 2012, 43–44). The problem is that that account hadn't been stress-tested until the rise of the Peasants across Western democracies.

For example, it had long been thought that "political leaders must trust one another in particular to give up power upon losing an election" (Lenard 2012,

43). Must? Tell that to The Donald. Trump's constant assault on American institutions has surely contributed to the record levels of mistrust in them, especially when institutional trust in other G7 nations is generally increasing (Saad 2023). Trump is not the only offender – Texas Governor Greg Abbott's reasoning for pardoning convicted murderer Daniel Perry, blaming virtually every level of Texan criminal justice, was nothing short of shameful (Hooks 2024). The disconnect also preceded Trump: Lenard cited fair electoral boundaries as important for trust (2012, 44), despite the fact that American districts have been gerrymandered to the point of caricature by elected politicians for years (Codrington 2022–23).

Yet even if trust is essential, the system somehow continues, plagued by mistrust, under threat, limping through. Even the Palestinian sympathisers who withheld their votes from the Democrats in 2024 were trying to have an effect by sending a message. They did not reject the system outright.

The populist Peasants are selective about whom they trust, rejecting democratic institutions (which they argue perpetuate oligarchy) and preferring to nurture an almost mystical link between the leader and "the people." They connect trust and legitimacy in a tight circle: "the re-entry of authorization under the guise of *trust* for the leader, and … an imaginary identification that suspends the distance between masses and authorities" (Arditi 2003, 23, his emphasis). As Trump put it on accepting the Republican nomination in 2016, "I am your voice" (Lukes 2019). But this risks giving unrestrained power to those who are trusted the most (Crick 1964, 63, citing Aristotle's critique of democracy), a structure suited to leadership of the mass man, "a type of man who does not want to give reasons or to be right, but simply shows himself resolved to impose his opinions" (Ortega y Gasset 1932, 73).

"Often the claim that we trust one particular legislator is based on more than the simple claim that she seems to have internalized the norms we think are valuable … . Rather … we are making a moral evaluation of a particular politician that is akin to the evaluation we make in … interpersonal trusting relations" (Lenard 2012, 49). This is true but incomplete: the Peasant trust complex combines the (strongly positive) moral evaluation of the leader with dogmatic mistrust of all other politicians. It may be that the opposition is over-intellectualised, lacks the popular touch, or is fragmented and incoherent. Either way, populists make it harder for them to get attention, and when they do get airtime they tend to blow it by repeating populist memes or appearing to criticise commonsensical or valuable norms. The strategy of the populist leader is to promote trust of herself by articulating important norms, while also creating mistrust by asserting that "politics as usual" is a sellout.

The leader's success is measured by the size of the mob of Peasants prepared to pick up their pitchforks. In terms of trust, it has three components. The leader must make a genuine connection; the threatened norms must be straightforward and plausibly shared, and arguably central to the success of a society. Hence, secondly, the leader must show that the political class has left behind or jettisoned moral legibility for the benefit of outsider groups which cannot plausibly be represented as having "created" or reproduced the valued norms across generations. She may argue that the political class's rejection of local values exhibits contempt for the people – which in many cases is trivial to do (Embery 2021). Third, she must represent the compromises and negotiations that are an essential part of getting-things-done politics as yet more betrayal of valued norms.

While the Peasant leader may be on strong ground with the first two components, the third is damaging to politics as a practical craft. Yet making that point invites the reply: "so much the worse for practical politics." And so the Peasants feed on their opposition, the more so if it is a remote, imperious, cerebral, cultivated clique devoted to process and detachment.

For the Puritans, it is axiomatic that non-Puritans are untrustworthy. Sociologist Manuel Castells began one work fretting about "malignant winds blowing on this blue planet … [a] maelstrom of multiple crises" including labour insecurity, low wages, terrorism, the existential threat of climate change, war, violence against women, post-truth and misinformation, and a culture which commercialises our demons. Not so good. But an "even deeper crisis, which has devastating consequences on the (in)capability of dealing with the multiple crises … [is] the rupture in the relationship between those who govern and the governed. The lack of trust in institutions, observed practically all over the world, delegitimizes political representation and so deprives us of a safe haven to protect us in the name of common interest" (2019, 3–4).

Certainly there have been serious pickles in recent years. Most obviously, there was the financial crash of 2007–09, where many resented the reckless lending and expansion of the banks leading up to the crisis, and the subsequent bailouts (which also allowed bankers to award themselves giant bonuses foolishly). In the UK, there was a major expenses scandal amongst MPs, some of whom went to prison. Some have argued that the public's patience has been worn down by constant policy failures, such as the inability to control immigration (Buckley 2018, 79–82; Timothy 2020, 138).

But while Castells was concerned about the decline of trust and the degradation of politics, he seemed also to think that the former was never high and the latter has never been out of the swamp. "Although Spanish politics is among the most corrupt in Europe, political corruption is in the basic DNA of almost any political system, the USA and EU included" (2019, 18). This is a long way from Crick's view, and spectacularly, almost irresponsibly false (Artés 2013). Spain was ranked joint 35th cleanest country in the world out of 180 in

2022 by Transparency International,[1] hardly brilliant but in the top quartile and nowhere near the worst in Europe. In the same ranking, the USA was 24th. Fifteen European countries ranked above Spain, including the United Kingdom (18th), and 11 countries in the EU. Castells highlighted Iceland as having corruption as a "systemic trait of contemporary politics" (2019, 18); it was joint 14th with Canada and others. Of the top 20, only Singapore (joint 5th) and Hong Kong (12th) were not liberal democracies. Eleven were constitutional monarchies, to which I say God save the King.

Castells, while concerned about the delegitimisation of representation, didn't seem to be an enthusiast for it, because of "the biased practices of members of parliament, judges and government officials." Democracy "is built around the very relationships of social power that founded it … privileging the power that is already crystallized within institutions." This has led to "the widely-held feeling that the agents of the political system do not represent us," which in turn means we can't even talk of representative democracy here (2019, 10).

On this nihilist view, Burkean representation crushes democratic aspirations because it "is limited to those who are already embedded in institutions and in society's vested interests, with all sorts of obstacles in the way for anyone who tries to gain entry to this well-defined political domain" (2019, 10). That would be news to Nigel Farage and Jeremy Corbyn in Britain, the German AfD and Die Linke, Emmanuel Macron, Marine Le Pen and Jean-Luc Mélenchon in France, Bernie Sanders and Donald Trump in the States, Pim Fortuyn and Geert Wilders in the Netherlands, Podemos, Ciudadanos and Vox in Spain, Syriza in Greece, the Five Star Movement and Giorgia Meloni in Italy, and so on. Certainly the establishment tends to pull together to exclude people such as these, and sometimes succeeds in pushing back, sometimes not. But with this rogues' gallery of insurgents, it's weird to harp on about the inability of representative democracy to upturn vested interests.

Castells had a particular down on political parties, complaining of their "distancing themselves from any oversight or decision-making by their members" (2019, 11), despite a long trend toward intra-party democracy. John F. Kennedy won the Democratic Party nomination in 1960 largely by mobilising activists in key states against party machine stitch-ups (White 1962, 78–80), creating the open Primary system that has handed the two major parties to their activists and more or less destroyed American politics as a constructive

[1] https://www.transparency.org/en/cpi/2022. To be fair to Castells, he was writing at a time when Spain was enmeshed in corruption scandals connected to the right wing Partido Popular government which he opposed. It had a lower ranking then, but not much – joint 42nd in 2017 (https://www.transparency.org/en/cpi/2017), just about in the top quartile. Furthermore, Spain's improvement gives the lie to the idea that it is uniquely and permanently corrupt.

problem-solving activity. British Conservative Party members got their first chance to elect its leader in 2001; they chose incompetent rebel Iain Duncan Smith over the respected and seasoned Kenneth Clarke. More recently, they chose Liz Truss over the MPs' favourite Rishi Sunak; she lasted an even shorter time than IDS. Party members may not make very good decisions, but they do appear to make them.

In the current chippy, anti-authority climate, many political commentators go over the top. The perennial danger is that they fail to grasp the benefits of a system that is no doubt flawed (as any functioning system in a complex polity is bound to be), and throw the baby out with the bathwater: "elements in such normally respectable doctrines as conservatism, liberalism, and socialism can show themselves unhappy, not just with any particular political order – as it is their proper, polemical business to be, but with political activity itself" (Crick 1964, 111).

The Mechanism Solution: Administrative Mechanisms

Representation, for thinkers such as Castells, is a dead duck. Free will is inimical to trust, so lawmakers must be bound (Castells 2019, 11). Let us call this the *mechanism solution* to the trust problem.

A mechanism must bind lawmakers at a point in time, either after their decisions, during decision-making, or in advance. Binding lawmakers *post hoc* is what we call *accountability*, and this won't do the trick, as it leaves their discretion intact. They can be voted out of office, but voters' memories can be short, or short term, and past failures with long-term effects might only come to light after the policymaker has left office (King & Crewe 2013; Dunt 2023). Binding them during decision-making is what we call *transparency*, and basically forces decision-making to take place in public. The problem with that is that it doesn't really enhance trust, and also inhibits the compromise and horse-trading that are essential ingredients of politics-as-getting-things-done (O'Neill 2002). Furthermore, if voters don't fully grasp the issues involved, then the details of decision-making will be just as baffling as the decisions themselves.

That leaves restraint in advance – the very means of decision-making, and the range of decisions to be made, must be restricted. "In the new political order … the decisions must be the result of scientific demonstrations totally independent of human will. … Under such an order we shall see the disappearance of the three main disadvantages of the present political system, that is, arbitrariness, incapacity, and intrigue" (Saint-Simon in 1821, quoted in Crick, 1964, 96). *A fortiori*, if decision-makers make no decisions at all, then they cannot make untrustworthy ones.

However, administrative mechanisms have not shown themselves to be solutions to these problems. Direct forms of democracy usually disappoint. The most familiar is a referendum on a controversial issue. Usually, these get called when the outcome is either predictable or manipulable, and they work best when a recognised part of a political tradition, as in Switzerland or the United States, and very badly when grafted onto a Parliamentary system where they present the legislature with a task to perform without willing the means (O'Hara 2006; and see the Introduction). In Britain, the Scottish Independence and Brexit referendums poisoned politics for years. Incredibly, "Immediately after the 2016 [Brexit] referendum, Nicola Sturgeon, leader of the SNP, started taking soundings about holding a second referendum on Scottish independence" (Clarke et al. 2017, 228). After opposing Brexit, on the (in my view justified) ground that dissolving a 43-year old political and economic union on the back of a momentary majority on a complex question reduced to an ill-defined yes/no binary was unwise to say the least, Sturgeon proposed to dissolve a 309-year old political and economic union on the back of a momentary majority on a complex question reduced to an ill-defined yes/no binary!

A variant on this is to turn elections into the equivalent of choosing a delegate and mandating a list of policies, rather as Burke's colleague urged in 1774. This doesn't seem like a terribly good idea, because decisions have to be made in a different context from that of election day (for example, healthcare post-COVID and defence policy post-Ukraine invasion might well be unhelpfully constrained by binding decisions made beforehand).

A stronger alternative is to vote on the policies themselves, as well as the candidates. The parallel universe below (from a British Liberal Democrat pamphlet), is intended to persuade us that 'distrust-based politics' is a good idea.

> The Alternative AV+ system empowers voters to choose their policy priorities in the sure knowledge that these choices will be reflected in the new parliament. Voters therefore do not mind spending up to fifteen minutes in the voting booth numbering their preferences across the three page ballot paper. Turnout reaches 90% as polling is spread across the Spring Bank Holiday every four years. Increasingly, voters use ICT facilities to examine the manifestos and research the options available before voting electronically during the three day election. (Tucker 1999, 51)

This political nerds' fantasy sounds ghastly. But even if practical, contradictory policies might be elicited – for instance, no new taxes, no excess borrowing, expansion of public services. Of course, contradictory policies emerge in representative politics too, but at least empowered representatives can do something about the problems when the faeces hit the flabellum, and may be held accountable at the ballot box next time round.

Another type of structure is the citizens' jury. This is another means of reducing representatives' discretion, or, from another perspective, allowing representatives to avoid responsibility for difficult decisions, so as not to fall victim to Juncker's dilemma. In a citizens' jury, ordinary people spend a few days discussing a tricky problem, and then – unencumbered by the need for re-election – make policy proposals, which may enable professional politicians to take them up. If they prove unpopular, the politicians have some cover, with the jury to blame (Pearse 2020).

They work like this: 100 or 200 people are chosen on some supposedly randomised basis, selected to be as diverse as you like, reflecting the national population distributed across gender, sexuality, age, ethnicity, able-bodiedness, education, employment, geographical location, or what have you. The only way they are not representative is the most important – 100% of them have nothing better to do. They are set up in a large hotel/conference centre with facilitators and a camera crew. Experts come and talk PowerPoint at them for a few days. The cameras roll, and they begin to debate and discuss. Sometimes someone says something with a bit of rhetorical dash, and that gets excerpted and spliced into the promotional video, with the music from *Chariots of Fire* playing in the background. They then produce recommendations, feeding back to the sponsor what the experts drummed into them at the beginning. The government receives this with great gratitude, and does what it always intended to do in the first place, quoting selectively from the recommendations. And finally, however close to the recommendations the final policy is, *The Guardian* finds a jury member to go on record that the government betrayed the whole process and never took it seriously, and so everyone is happy. For a more sophisticated critique, see (Tucker 2008).

Another method is to insert specific policies into the constitution, thus making them harder to countermand. This micromanagement of the future never – never! – works. Making abortion a constitutional right poisoned American politics for forty years, and since the repeal of *Roe v Wade* abortion's proponents find themselves in a far better position than ever they were while it was in effect (Kamarck 2024). The constitutional balanced-budget measures (the "debt brake") in its constitution have prevented Germany from addressing some serious issues, although they are popular among the public, especially those who don't trust politicians (Hayo & Neumaier 2016). The Brazilian constitution has a federal spending cap too, as well as several expensive commitments (e.g. on pensions) that should be matters for Congress. No wonder it has been changed well over 100 times since it was drafted in 1988, largely to allow increased borrowing (change to support political reform is far harder to get through). Strongmen everywhere like to give themselves more power, or to amend term limits, if it is easy to change the constitution; if not they ignore it, even when they have written it themselves.

The most absurd constitutional farrago of recent years occurred in Chile between 2019–23, the most successful country in Latin America (Zitelmann

2019, 92–107). In a climate of protest, political carping and navel-gazing it was decided to change the Pinochet-era constitution; the Puritans took over the process and put in ludicrous rights for the promotion of Chilean cuisine, and for children to be taught empathy for animals. It was rejected in a referendum (the good sense of the Chilean people on show). The Peasants hijacked the next draft, and wrote in their own obsessions, such as abolition of a property tax and a right for business owners to refuse services to gay people. It was rejected in a referendum (the good sense etc., etc.). The result is a population that is browned off with politics, and still saddled with the original constitution.

The Mechanism Solution: Technocratic Mechanisms

Administrative mechanisms are too indeterminate. This suggests that Saint-Simon's "scientific demonstrations totally independent of human will" require some technology to make them happen. Certainly, untrustworthy decisions can be bettered by sensibly-designed tech. Artificial intelligence, for example, is likely to produce fairer decisions, even on biased data, than a biased person, and it is probably easier to remove the bias from the data than from the person (Lobel 2022).

The Puritans have a special affinity with technocratic anti-politics, rational policy, effective administration, targets, accountability, disinterested judgment, in other words, pushing people around for their own good. More critical Puritans have wondered whether disinterest is possible; Castells was torn between admiration for the "competence and goodwill" of mandarins, and concern about their legitimacy, because they are not chosen by the people but rather imposed from above (2019, 80). The circle he would need to square is that if officials were elected, their work would be politicised, and the competent and good may not (probably would not) be elected.

Moreover, civil servants are people and have agendas, so as technocrats they may be trusted more than they ought to be (Crick 1964, 109). Hobbes' Leviathan solved problems of trust and trustworthiness in the state of nature through its ability to constrain the population, but the modern state is a large organisation made up of individuals and agencies competing for resources, and so Leviathan itself needs something to constrain it, and it is not hard to detect the beginning of a regress (Hollis 1998, 32–33). Furthermore, we are where we are, and not in the state of nature.

Technocracy affects Puritan/Peasant dynamics. The apparently polar opposites of populism and technocracy are "trapped in a mutually reinforcing dynamic" (Rummens 2017, 567). Populists try to repoliticise those areas of society that have been technocratically depoliticised. Meanwhile, populist pressure, and outright successes, convince the technocrats of the flaky nature of democracy, which leads them to try to insulate more key policy areas from competitive politics.

Yet despite their antagonism, both technocracy and populism reject the practice of getting-things-done politics in a plural society. They carry parallel critiques of democracy, intermediation and faction (Bickerton & Invernizzi Accetti 2017; O'Hara 2017a). The idea that interests should aggregate and express themselves through political parties or civil society associations goes against the populist idea that the will of a unified people can be expressed by a leader. Meanwhile, technocrats see interest-based politics as a quest for undue influence, diverting from the higher rationality of their mission.

They also jointly reject procedural legitimacy, where legitimate outcomes emerge from following the rules. For the populist, the will of the people is what counts, and the checks and balances of the normal liberal constitution are impediments to it. For the technocrat, expertise is the criterion for legitimacy, and when democratic procedures deliver a different outcome, they cannot be legitimate. Some technocrats piously hope that expertise and the general will can be married without a shotgun: "We need to posit a collectively controlled legitimate vertical authority in addition to horizontal forms of sociality, to avoid becoming the slaves of either a tyrannical totalitarian centralism or a capricious emergent order beyond our control" (Williams & Srnicek 2019, 359) – but where there is a clash, the experts must take precedence to prevent the emergent order getting *beyond our control*. The referendum shall be repeated until the people get it right.

LEVERAGING THE NETWORK SOCIETY

Both Peasants and Puritans have an instinctive dislike of hierarchies (outmoded), and are attracted by networks (fashionable). Hierarchies comprise agents connected vertically, where authority descends from above, while networks are horizontally connected, so that everyone is on the same level. The latter appeals to both sides, given their dislike of existing elites. The Peasants prefer a very flat hierarchy, with a charismatic leader at the top, trusted by everyone below, but so sparse are the hierarchical relations otherwise that in most ways it behaves very like a network.

These structures have trust implications. Vertical trust connections, where there is an asymmetry in power or authority between trustor and trustee, are generally speaking characteristic of areas where authority and expertise are exercised, such as science and politics (O'Hara 2004, 91–92). Many social scientists have argued that vertical structures rely on power rather than persuasion, and so are problematic for trust, which "is a fundamentally egalitarian ideal, … [to which] hierarchy is inimical" (Uslaner 2002, 3).

Networks are resilient in low-trust environments. Even if each network member trusts only a few of his fellows, it can still stay connected. In a hierarchy, on the other hand, trust needs to travel in abundance up from the many at the bottom to those with authority at the top. Failing trust will isolate those at the top, so that their orders or advice are disregarded at the bottom. Even when there is trust enough to sustain a hierarchy, bottlenecks can develop in the flow of information; those at the bottom may have to sit waiting for overworked authorities to get round to their problems. On the other hand, networks scale easily, as there is no central point to become a bottleneck. The relative advantages of networks over hierarchies through history and in business and technology have been widely rehearsed in a series of studies (Raymond 2001; Ferguson 2017). The main advantage of hierarchies is that they are easier to direct, for example to perform a task. In a hierarchy, a decision is taken by someone in authority, whereas a network has to achieve consensus. The flat hierarchies of the Peasants are designed to get the best of both worlds.

As if on cue, the new information technologies of the Internet and the World Wide Web facilitated the rise of the "network society" at the end of the 20th century.

> Networks constitute the new social morphology of our societies, and the diffusion of networking logic substantially modifies the operation and outcomes in processes of production, experience, power and culture. While the networking form of social organization has existed in other times and spaces, the new information technology paradigm provides the material basis for its pervasive expansion throughout the entire social structure. Furthermore, I would argue that … the power of flows takes precedence over the flows of power. (Castells 2000, 500).

The Internet, Web, smartphones, big data, cloud computing, artificial intelligence and machine learning, platforms, surveillance capitalism, streaming, text, image, video and voice processing, drones, telemedicine and many other interoperable technologies reinforce each other in an ambient data infrastructure which has cemented the network society in place in digital modernity. The two visions of technocracy and radical democracy have been able to coincide, sometimes uneasily, as the new paradigm of digital modernity valorises *both* the control that data analysis and feedback can provide over real-world environments, *and* the relentless disruption associated with constant innovation (O'Hara 2020; O'Hara 2022a). Burkean representation, in comparison, looks like an antiquated hierarchy.

As we saw in the previous section, older mechanisms for addressing the trust problem are deficient. There are regulations, social norms and financial incentives, but one is always free to break the law, act abnormally, or eschew material gain. Technology is different – it creates a palette of options, but closes off what it doesn't make available (Lessig 1999). The forbidden option on the menu, as it were, is greyed out.

A Dead End: Horizontal Politics and Protest

Radical thinkers have welcomed the network society as liberation from the humiliating requirement to trust leaders and representatives. There are no rules except freedom of expression and personal agency, and imagination is unconstrained (Castells 2019, 20–21). "In this situation of structural crisis and institutional lockdown, it is autonomous social movements, as so many times before in history, that are exploring new forms of collective action rooted in the real lives of the people who produce, live, suffer, love and make projects within the fabric of the human experience" (Castells 2019, 86).

The idealism of "horizontal politics" can be illustrated by the Occupy movement, which emerged as a furious response to the financial crisis of 2007–09 (Pickerill & Krinsky 2012; van Stekelenburg 2012; Gamson & Sifry 2013). One of the organisers of Occupy Wall Street drew a direct connection to the affordances of networked technology: "the interconnections of communication are such that you can imagine people not just communicating but acting, and acting damn effectively, without leadership, a secretariat, without even formal information channels. It's a little like ants meeting in an ant-heap, all waving their antennae at each other, and information just gets around – even though there's no chain of command or even hierarchical information structure. Of course it would be impossible without the Internet" (Graeber 2009, 32).

The trouble is that lack of leadership tends to produce a reactive group, at best passive, most likely self-indulgent, at worst squabbling. Politics where principle is pursued to the exclusion of everything else doesn't intersect with the real world, and would be as harmless and pointless as student politics, if it didn't offer people false hopes (Crick 1964, 134). Yet some are satisfied with student politics. Judith Butler even argued that it was important for Occupy to make nebulous, unsatisfiable appeals because "articulating demands that can be satisfied depends fundamentally on the attribution of legitimacy to those who have the power to satisfy the demands" (Butler 2012). This odd claim made a virtue of a necessity, because the uncoordinated groups that made up Occupy couldn't have agreed on a concrete, achievable set of goals if they occupied till doomsday (Roberts 2012).

As an example of the naïve and gushing gibberish that drove not only policy but method, consider this paean to the Occupy Wall Street drum circle:

> The music was often heard by white American publics as though belonging to a distant primitive past and a distant place outside the limits of the cosmopolitan city. This otherness was part of its appeal for many bohemians, dropouts, and

> counter-culturalists who sought to reject the values and norms promoted by mainstream society, and hand drumming became a means to signify a belonging to its margins. … The drum circle at OWS was a site of intense joy and conflict, its relentless polyrhythms filling the camp and reverberating throughout the Financial District's thicket of glass towers. (Mindel 2021)

Actually, the heartwarming "Afro-modernity" drove everybody bananas, and the drum circles were asked to stick to office hours. As the protest continued, the East and West sides of the protest camp somehow became antagonists in a *Lord of the Flies* sort of way, while violence and drug dealing proliferated (Roberts 2012, 757).

The Spanish *indignados*, or angry ones, were more successful, spawning a political party, Podemos, that achieved some electoral success (Castells 2019, 104–111). This could only be done at the cost of replacing its horizontal network structure with a flat populist hierarchy, with pony-tailed Pablo Iglesias at its head. He described his rise and the party's significance thus:

> The 'People of the Television' … didn't know about a new political party called Podemos, but they knew about the guy with the pony-tail. This populace, politically socialized through television, was not 'representable' within the traditional left-right categories of the political space. In the context of high dissatisfaction with the elites, our objective of identifying a new 'we' that included the tv nation initially came together around the signifier 'Pablo Iglesias. (Iglesias 2015, 17)

This splendidly anticipated Trump's characteristically egotistic yet flattering wooing of the Republican Party: not "I am your voice," but "I am your signifier." Exquisitely hip! Podemos had its ups and downs in the intervening years, at one point joining a government coalition. Iglesias himself left politics in 2021, and the party retreated to the fringe.

Horizontalism was a dead end, but it used a technology – the Internet – as an image for a particular ideal of social connection. Coming down the line was another technology that could be an even more powerful metaphor of how to move beyond student politics without requiring a leader.

The Other Product of the Financial Crisis: Bitcoin

Even before Occupy collapsed into self-indulgent chaos, a technology was appearing in response to the financial crisis that would serve as a better metaphor for the trust-free mechanisms that Puritans and Peasants wanted to create: *blockchain*.

A prosaic description of blockchain is a peer-to-peer technology (i.e. it's a decentralised network in which everyone has equal status) for recording associations between two things, such as property and its owner (i.e. it's a *ledger*).

In the hype it's an evolutionary step in the history of institutions (Davidson et al. 2018). It was originally proposed by a mysterious, possibly non-existent person called "Satoshi Nakamoto" as an architecture to underpin a digital currency, Bitcoin (Nakamoto 2008).

Nakamoto's invention is extraordinarily brilliant. Bitcoin is a currency (a "cryptocurrency") that is totally different from those issued by states ("fiat currencies"). It has no administrator to decide when and how to issue it, and so it can't be inflated by governments to buy their way out of a temporary hole. Instead it is produced by a network of "miners" who jointly create and store a transparent cryptographically-secured record of all Bitcoin transactions – the blockchain (O'Hara 2015).

Traditional, hierarchically organised banking requires a hefty infrastructure of administration, regulation, auditing, identification systems, cybersecurity and enforcement to support trust in the system, all of which make expensive demands for compliance, provide opportunities for consultants and allow governments to change the rules to suit themselves. The claim of blockchain is that trust can instead be provided by a public ledger with transparent records; anyone with concerns can simply inspect the chain. A blockchain has no central location; it lives on those nodes in its network which choose to host the complete ledger (hence its posher name of a *distributed ledger*), and so no-one can be denied access. The only way that it can be altered is by the peers in the network themselves, only as long as a suitable majority of them agrees. In a large network, this should be hard to achieve (though see Chapter 9).

A blockchain is simply a database. In Bitcoin's case, it contains two types of information: complete records of (a) the creation of each bitcoin; and (b) each transfer of bitcoin between users. Until a transaction appears on the chain, it has no verifiable existence; once it has appeared, it is there forever, for all to see. As it is a database, it can store information about the connections between anything, and need not be focused on money. The information stored can be a computer program, which then might be executed on the blockchain without human intervention: inexorable, transparent and verifiable, the ultimate in mechanism solutions. Where such programs have been set up to perform a transaction and agreed by the various parties, they can't be forestalled, and so have been called *smart contracts*, to which we will return in Chapter 9.

It therefore – in theory – does away with the need for trusted intermediaries, regulators and bankers who can always go back on their word (Steinmetz et al. 2020, 115–171). Even those without an ideological gripe worry that "'intermediaries of trust' are slow to respond to changes in the environment, add cost and time to transactions, and adversely affect productivity" (Khurshid 2020).

The Blockchain World of Trustless Trust

The blockchain is explicitly designed to enact a particular type of mechanism solution, which has been called, oxymoronically, *trustless trust* (Kiviat 2015, 574; Bahga & Madisetti 2016, 543; Klems et al. 2017). No-one in the network need trust anyone particular in the network, but it is structured so that the outcomes are trustworthy. There is no easy and secret way for any individual to influence decision-making, because the ledger is transparent and open for inspection.

Why an oxymoron? Well, it's not that trust has disappeared, but its object has shifted from intermediaries to the technology (Gilder 2018, 242; Chohan 2019; Waldo 2019; De Filippi et al. 2020). It does not manufacture trust, and it is not a "trust machine" (*pace* Economist 2015). Trustless trust taken literally sends us into an infinite regress. Some sort of trusted baseline or infrastructure is required – it might be technological, or it may involve trust in the crowd, the many over the few. Transparency enables each action to be observed and disciplined, but we have to trust that the blockchain correctly represents something tangible.

Let us define trustless trust, then, as this: a means for furnishing confidence in another's future behaviour, by deliberately reducing their degree of freedom, perhaps through incentives, mechanisms or regulation, ideally close to zero.

Oxymoronic or not, it appeals to both Peasant and Puritan, libertarians who valorise individuals' resistance to oppressive social forces (Land 2019; Tate 2019), and those of an anarchist bent who oppose the state and capitalism, and prefer strong, uncoerced, bottom up mutual aid (Williams & Srnicek 2019). The individual may trust only his own small community, or indeed no-one else at all, while living an authentic life doing his own thing.

Today's systems-oriented digital modernity provides few alternatives for resistance, which is experienced by many as oppressive and iniquitous. Some signal their disregard by (on the left) going "off-grid" (Forde 2017) or (on the right) becoming "survivalists" (Jackson 2018) or "preppers" (Mills 2019), but it is of course impossible to detach oneself from the corruption of civilisation while simultaneously using digital technology. At a minimum, you have to pay your electricity bill. Blockchain has therefore been proposed as the ideal technology for working through possible futures with trustless trust, whether crypto-libertarian or rooted in a commons (De Filippi & Leiter 2021). Some discuss "crypto-secession" (Allen et al. 2018), opting out of government services using smart contracts, cryptocurrencies, small-scale renewable power, self-sovereign identity systems and decentralised management to ensure equitable provision of public goods, with a transparent ledger of what has been provided to whom.

BLOCKCHAIN POLITICS

My purpose in this book is not to explore the direct potential of blockchain – many others, usually enthusiasts, have done that (Economist 2015; Allen et al. 2018; Gilder 2018; Steinmetz et al. 2020). Personally, I am sceptical that blockchain can deliver the data integrity that is the foundation of its promise, and doubt that it will be as transformative as its boosters claim, but that isn't where I am going here. Indeed, I am supportive of the measured use of a brilliant technology for creative purposes, particularly within large organisations or associations in dynamic contexts where data-sharing needs to be accurate and timely.

Rather, I want to look at political trust. When I write of *blockchain politics*, I connote a *style* of politics based around trustless trust as defined above, not the literal use of blockchain for political purposes.

Blockchain politics brings together three distinct tenets, critiques of the practice of trust.

- First, *centralised institutions or intermediaries can never be assumed to be trustworthy*, because those within them are in such privileged positions they have no incentive to be trustworthy. Elites will use the system to reinforce their domination. Intermediaries centralise to reap economies of scale, so disintermediating "provides opportunities for a more localized and individualized society" (Steinmetz et al. 2020, 100). Intermediaries (including Burkean political representatives) are rent-seekers; it is not in their interests to add more value for the principals in the transaction than they cream off in fees.
- Second, unless the stakes are low, *people can never afford to trust each other.* Trust exposes those who are competing, or whose interests conflict or diverge, to needless risk. Self-interest will lead to collective action problems, free riding and tragedies of the commons. Those whose interests do not coincide will never look out for each other's welfare, so representation can never be satisfactory unless: *either* one is represented by someone with identical interests; *or* the representative's freedom to use her discretion is sharply reduced.
- Third, *decentralisation, mechanism and transparency are the only reliable means of supplying trust between strangers.* It is only by relentless scrutiny (on one's own behalf, that is – not delegating the job to an intermediary) that one can be assured that others are doing what they said they would. Since the purpose of representative democracy is to remove citizens from active politics, it follows that it exposes them to excessive risk.

In sum, *trust is a vulnerability* the need for which should be eliminated. Trustless trust is the only way to renew confidence in the political process of those who trust neither authorities nor their fellow citizens. In a sense, this is done by removing the possibility of politics altogether (Crick 1964, 148–149). The faith in institutions and systems characteristic of modernity (Giddens 1990, 79–111) is leveraged by the creation of mechanisms, whether technical or political, to reduce discretion.

One caveat: the Peasants do have a need for trust, because typically someone leads them. The pitchfork mob is a very flat hierarchy – a leader and everyone else. There is no room for a competitor – the Cruzes and DeSantises can only get so far before they face a three-way choice: reverse and show obeisance to the Trumpian numen, bring off a *coup*, or leave the group entirely. This structure does require trust, admittedly, but of a very specific kind, in a single direction and of one person. It can therefore easily be withdrawn in the event of perceived betrayal, but this tends not to happen. Negative evidence is frequently outweighed by the deep faith that the leader is able to conjure up, reinforced by the peer pressure, mutual respect, status, and belonging that tight-knit group dynamics provide (Obradović et al. 2020).

Is the world of trustless trust possible? Or desirable? Those are the questions behind the inquiries of Parts I, II and III.

PART I

A theory of trust

Introduction

The first task is to sharpen our understanding of trust and trustworthiness. Although much has been written about trust as a fundamental constituent of the modern world, the "dark matter" that holds society together (Dunn 1984, 287; Hollis 1998, 1, 10; O'Hara 2004, 51–73; Lenard 2012, 9; Urbinati 2015, 63; Luhmann 2017, 5; Rotenberg 2018, 2–3; or, for Uslaner, "chicken soup" – 2002, 1), there isn't an overwhelming consensus on what it is (appropriate for the "dark matter" imagery). My own account isn't particularly outré, but the details will be important for the conclusions I draw from it in Parts II and III.

In one sense, it is relatively easy to specify: it is an affirmation of the trustworthiness of another (occasionally, oneself). Trustworthiness is therefore prior. This may appear circular, but isn't, as trustworthiness is not defined in terms of trust (it will be discussed in detail in Chapter 3). Trustworthiness is a property of an agent. Trust is an attitude that one agent has toward another (specifically, a judgment, instinct, opinion or appraisal of the trustworthiness of the other). It is intentional, a cognitive state that relates to another.

Trust is often reciprocal – you and I trust each other to be trustworthy (possibly in different ways – I trust you to provide a good service, and you trust me to pay the fee). It is hard to disentangle such reciprocity, and it is not clear that mutual trust decomposes exactly into my trusting you and your trusting me without residuum. However, I will take "A is trustworthy" and "B trusts A" as the key predicates to be defined. More complexity might have to be added for a complete account, but the aim of Part I is only to furnish enough to support the later argument.

Niklas Luhmann's account of trust's functions has been justly influential (2017, 27–35). Certainly its serving these functions gives it important utilitarian value, but the phenomenon of trust doesn't seem *primarily* functional; we don't often judge whether trusting will maximise well-being before placing or withdrawing it. Many, I would say most, cases of trusting and being trustworthy are pre-reflective, instinctive, even habitual.

"The problem of trust" is often described as the need to increase trust, but it is clear even from this short discussion that this is not accurate. It is no advantage to trust someone who is not trustworthy, so if trust is to be increased, we need to make sure that trustworthiness keeps pace. The *real* problem of trust is *how to align trust and trustworthiness*, so that, as far as possible, we trust all and only trustworthy people. *All*, to maximise interaction and overlook no valuable opportunities for collaboration; *only*, to avoid fraud. Yet this again is too utilitarian for a complete account. The preservation of trust and trustworthiness, or at least conditions conducive to them, is essential to the maintenance of what Burke called little platoons, beneficial associations and communities, and the smooth running of the state. Society depends on the *reflex* of trust, the uncomplicated acceptance of the obligations of one social being to another. The implications of this will be explored in Part II.

Where there is trust, we talk of the trustor (who trusts), and the trustee (who is trusted). This implies that "trust" is binary, and a success word – it has either happened or it hasn't. If it hasn't materialised, for whatever reason, we don't really have words for would-be trustors and trustees who aren't trusted. To make discussion easier, I will introduce the characters of Alice and Bob (and a few others along the way). Unless stated otherwise, "Alice" will refer to a potential object of trust, a trustee or would-be trustee, and "Bob" to a potential subject, a trustor or would-be trustor.[1] The aim of Part I is to outline the meaning of "Bob trusts Alice" sufficiently to inform the discussion of political trust in Parts II and III.

Alice and Bob are helpful in the deployment of pronouns, since they attract different ones. This will enable the use of anaphora in complex sentences, while still disambiguating between trustor and trustee (e.g. he trusted her to give him her papers before she went on her business trip). I will carry this convention through the book, so that masculine pronouns will, in general, be used to refer to the *subjects* of the trust relation, and feminine pronouns will, in general, be used for its *objects*. It should go without saying that both these sets of pronouns are used in a general and neuter spirit; that is, I do not claim that trustors or would-be trustors are exclusively, chiefly, ideally or generally boys or men, or that trustees or would-be trustees are exclusively, chiefly, ideally or generally girls or women.[2]

---

1 The names "Alice" and "Bob" are borrowed from a standard scenario in cryptography (https://en.wikipedia.org/wiki/Alice_and_Bob). I associate Alice with the prior notion of trustworthiness only because "Alice" comes first in alphabetical order.

2 Pronouns are something of a minefield. A blurb for a recent book on them asserts that they are "one of the most pressing cultural questions of our generation," which they most certainly are not. I have set out the practical rationale for my pronoun use in this book, and I intend grammatical gender to do some serious work of disambiguation. If that gives you the vapours, then I hope you can take a deep breath and carry on regardless. If you cannot, then you are probably one of the targets of the argument, and should read it anyway.

Trust usually accompanies uncertainty. Hence, while Bob's attitude to Alice may rest on some knowledge he has of her, his trust itself is not knowledge. If trust is not a *sui generis* mental state, it is a *belief* on his part; he *believes* something of Alice that falls short of knowledge about her (I share this view with, e.g. Seligman 1997, 43). I suggest the most obvious thing that he believes is *that she is trustworthy*. If he trusts Alice, he surely believes that, and conversely, if he believes her trustworthy, what stands in the way of him trusting her in some sense? Hence the trustor believes that the trustee is trustworthy. It does not follow from that that Alice *is* trustworthy, only that Bob believes her so (although in itself that may be valuable evidence that she is). There are many conditions, parameters and exceptions to this, which will be filled out in Chapters 3 and 4. Chapter 3 will look in some detail at the implications of saying that "Alice is trustworthy," and Chapter 4 will go on to consider "Bob trusts Alice." Before we get onto these, however, further preliminaries will be discussed in Chapter 2.

2. Preliminaries

The term "trust" is used ambiguously in both ordinary and academic discourse, a point which many have noted but few have unpicked in detail. In this chapter I will look at why this is the case, and at some specific ambiguities which can cause misunderstandings. I will also briefly consider the morality of trust and trustworthiness, and consider some arguments about the range of their historical application. This is something of a scoping exercise, prior to more detailed discussion in Chapters 3 and 4.

FAMILY RESEMBLANCE

There are multiple ambiguities in the use of the term "trust," a phenomenon which suggests the relevance of Ludwig Wittgenstein's family resemblance theory of meaning (1958, §§65–71).[1] This describes terms not with absolutely separate senses, but that connect different but related conceptions of a wider cluster of ideas (O'Hara 2023, 85–102). There are no necessary and sufficient conditions for trust, but rather resemblances of varying significance between those relations that we see as trusting; "you will not see something that is common to all, but similarities, relationships, and a whole series of them at that … a complicated network of similarities overlapping and criss-crossing: sometimes overall similarities, sometimes similarities of detail" (Wittgenstein 1958, §66). The similarities exhibit *family resemblance.*

This is deliberately abstract, and when we drill down into more concrete contexts, we find all the ambiguities we would expect from a family resemblance term. We can always resolve ambiguity in context by being specific, at the cost of wider applicability. Family resemblance terms create this dilemma: precise definitions exclude too many clear cases, while those definitions which cover the full range of cases are too vague to be of analytical value.

[1] I prefer Anscombe's 1953/58 translation of Wittgenstein's *Investigations* to Hacker and Schulte's revision of 2009 on aesthetic grounds. Given that I am not doing Wittgenstein exegesis, I quote the former. However, where Hacker and Schulte do improve on Anscombe is to detach Part II of the *Investigations* from the main body, and I will cite their version of 'Philosophy of psychology – a fragment' where I reference it.

Forms of Life

Let us consider context. Human behaviour, and the language that evokes it, takes place within meaningful social contexts, what Wittgenstein called "forms of life," whose occupants have a shared understanding of action and interaction. At their most formal, forms of life are role- and rule-governed, but they may be very flexible in more vernacular contexts. Those engaged and experienced in a form of life are reliable within it, and their interpretations of it tend to coincide. It constrains the language used, while also rendering it meaningful (within what Wittgenstein called a "language game"). Our understanding of institutions, conventions, practices, norms and other forms helps render someone's activities understandable, explainable and predictable (Wittgenstein 1958, §19; Baker & Hacker 1985, 238–243; Boncompagni 2022). Forms of life might include travelling on buses, attending conferences, being friends, playing tennis, debating, partying, or taking part in a court of law. They are unspecific about exactly what needs to be done, but there is tacit agreement on how to behave, what the limits are, and what (if anything) is to be achieved by the activity. Departure from the conventional conduct of a form of life is surprising, and often the subject of surreal comedy.

Unlike some commentators, I don't think Wittgenstein meant anything precise by "form of life" and "language game," but rather wanted to highlight the connection between what we say, what we do, and the social context from which these gain meaning and through which they can be effective. A common sociocultural background, suitable for the biological organisms we are, is necessary for meaningful communication and practice. Trusting and being trustworthy are both forms of life, and also components of forms of life.

Wittgenstein's informality is helpful. Suppose Bob gets on a bus. For trust theorists who base their accounts on bounded rationality, this looks problematic: as he gets on the bus, he must trust everyone on it not to do him harm, and yet he has no idea exactly who is on board. Hence he appears to be taking a reckless risk, but as we know, people get on buses all the time, and some of them even survive. But if we think of trusting as a type of behaviour embedded in a range of forms of life, then the appropriacy of foregrounding risk diminishes.

One may be in a situation where explicit trust judgment is simply inappropriate. Deciding whether to trust is a kind of doubt, which the form of life or language game may not support: "the absence of doubt belongs to the essence of the language-game, … the question 'How do I know …' drags out the language-game, or else does away with it" (Wittgenstein 1979, §370). Bob doesn't make a *decision* at the bus-stop to trust everyone based on zero evidence; rather, the whole idea of travelling on the bus *depends* on his parking his trust issues and simply getting on. This is not because it is not possible for him to be mistaken that there is no reckless murderer on board (that does happen, and where transport systems are known to be dangerous, it may be sensible to talk of trust judgments being made – Smith & Clarke 2000). Yet in normal circumstances, trust is a side issue: the process of being a bus traveller *involves* the adoption of an attitude of total confidence that one is safe. "Why do I not satisfy myself that I have two feet when I want to get up from a chair? There is no why. I simply don't. This is how I act. … Must I not begin to trust somewhere? That is to say: somewhere I must begin by not-doubting; and that is not, so to speak, hasty but excusable: it is part of judging" (Wittgenstein 1979, §§148, 150). And one *is* safe; on the bus, what René Girard called "the human genius for discord" (2023, 4) asserts itself, not in murder, but rather in behaviour such as inappropriately seeking conversation, or littering, or playing loud music. And even that is rare.

"This game proves its worth. That may be the cause of its being played, but it is not the ground" (Wittgenstein 1979, §474); "a language-game does not have its origin in *consideration.* Consideration is part of a language-game" (Wittgenstein 1981, §391, his emphasis). Wittgenstein drew attention to reflexes; Bob "plays the part" of a trustor instinctively, as it were, because that's what people who travel on buses do. Were he genuinely concerned about his safety, then he would not be a bus traveller in the routine sense. Doubt is for people who are not on the bus, including philosophers and sociologists reflecting on trust. Even they might take the bus home from their campuses, despite not having solved the problem of the rationality of Bob's trust. Once on the bus, they transform themselves from researchers to bus passengers, and a different set of attitudes is in play.

Doubt is incompatible with most quotidian activities. Those who do find themselves having to make inappropriate trust judgments are badly handicapped. In the psychotic condition of hyperreflexivity, a symptom of schizophrenia, patients over-analyse, worrying about questions whose answers are normally assumed: why did that person say "hello?" Are the bus passengers likely to attack? They focus intensively on detail, missing the basic *Gestalt* which intersubjectively connects socialised people: "what is taken for granted suddenly becomes unfamiliar or strange; what was implicit becomes explicit and enters the focus of attention" (Fuchs 2010, 239).

In the abstract, trust and trustworthiness are mysterious. Embedded in a form of life, however, we see them for what they (often) are: reflexes based on the simple and largely unconscious acquiescence in obligations that may be instinctive and certainly learned at an early age. Failing to doubt others brings with it vulnerability, but that is a by-product of the form of life, not of trust. I have focused on trust in this section, but much the same goes for trustworthiness – when Alice engages in bus travel, part of that involves her leaving behind any desire she may have to attack fellow passengers.

Ultimately, a person learns to accept "what has to be accepted, the given" in his current social *milieu* (Wittgenstein 2009, xi §345; Wittgenstein 1980, §630). The given is not ineffable, unstructured experience, but rather the interpreted cultural situation augmented by relevant conceptual structures, forms of representation, social norms, and in particular the extent of the autonomy of different agents which leads ultimately to the problem of trust in the first place.

Where the Resemblance Gives Out

As well as the family resemblances between examples of trust and trustworthiness, there are similar practices which are distinct enough to be classified differently, but which may be conflated. Some have argued that trusting is like accepting a promise (Hawley 2019, 27–47), or that being trustworthy means that, *inter alia*, one reliably keeps one's promises (Rotenberg 2010). Annette Baier argued, wrongly, that promises are invitations to trust, more convincing than just saying "trust me!"

> Part of what makes promises the special thing they are, and the philosophically intriguing thing they are, is that we *can* at will accept *this* sort of invitation to trust, whereas in general we cannot trust at will. Promises are puzzling because they seem to have the power, by verbal magic, to initiate real voluntary short-term trusting. They not merely create obligations apparently at the will of the obligated, but they create trust at the will of the truster. (Baier 1994e, 111, her emphasis)

This cannot be correct, because a promise, even when accepted, creates no trust. The value of a promise *depends* on how much the promisee trusts the promisor, so the required trust has to be already in place. The promise merely activates it. Suppose Bob is arriving at the station at 5.00, and has accepted Alice's promise to give him a lift. She is thereby obligated to be there at 5.00. If Bob (antecedently) trusts her, he will take her promise at face value, and wait for her at the station – if she is a few minutes late, no harm done. If she breaks her promise and proves untrustworthy, he may be upset. On the other hand, if he does not trust her timekeeping or her conscientiousness, he will make sure he arrives at the station armed with the numbers of a couple of taxi firms, *even though he has accepted her promise*. He might give her 15 minutes to turn up, and if not, he will call a cab. If she turns up, all well and good. In this second case, Bob accepts Alice's promise, and is prepared to give her the opportunity to carry out her obligations, but because he does not trust her very much, he does not rely on her, and comes with a plan B. Alice's promise has *no* power to initiate voluntary trusting, but rather is a commitment that Bob judges *through the lens of his (antecedent) trust for her.* At best, the promise gives Alice a platform for demonstrating her trustworthiness.

Contracts are similar, where the commitments are more formally set out and backed by a body of law, giving the parties sanctions or constraints that they can apply *in extremis*. Contracts will be discussed in more detail in Chapter 9, but for now I draw attention to some essential differences from trust. First, as with promises, trust precedes the signing of the contract. For most business arrangements, the parties sign if/when they trust each other, and if they don't trust each other, they may well not sign. Alternatively, a contract, in binding the trustee, may signal a *lack* of trust. A prenuptial agreement between two affianced people (Margulies 2003) may be realistic, but hardly romantic. In a trusting relationship, the parties may go beyond the letter of the contract to benefit each other, but where there is a lack of trust, signatories can interpret the wording mischievously (as with Shylock's contract with Antonio in *The Merchant of Venice*).

Furthermore, contracts stabilise things that trust leaves indeterminate. Contracts may be broken without comeback, as long as the breach is compensated, whereas betrayed trust may not recover even with generous compensation. Contracts are voluntary agreements between (usually) nonintimate equals, whereas many trusting relationships are personal or embody power asymmetries, such as doctor–patient and parent–child (Baier 1994e, 114–120).

Trust, contract and promising clearly all operate in the same region, but the first can be seen to be more fundamental than, and therefore distinct from, the others.

SOME AMBIGUITIES OF TRUST

Because trust is not a unique product of a specific type of society, we find the term being used in a number of different but related ways. In this section, I will sketch some of the ambiguities that are part of the complex swirl of family resemblances in our use of it and related terms.

Trust and Reliance

The paradigm case of trust is that of one person trusting another, Bob trusting Alice. However, the object of trust need not be a person; it could be a group (Bob trusts socialists), an organisation (Bob trusts the Salvation Army), a system (Bob trusts banks to manage his money), a technology (Bob trusts the aeroplane not to crash), or an object (Bob trusts the cliff not to collapse into the sea). Trusting expert systems (Giddens 1990; Luhmann 2017) is characteristic of our system-oriented world.

It may stretch the meaning of "trustworthy" to apply it to systems, buildings, bridges, procedures, artifacts and so on. Hence there is a simpler sense of trust, the belief in something's *reliability*, that its performance will be predictably up to a standard. Unlike a person, a system or artifact doesn't have a choice of whether to be reliable; it either is or it isn't. Belief in reliability and belief in trustworthiness are nevertheless related, in that they both suggest certain expectations on the part of the trustor, a view expressed by the splendidly-named Julian Rotter (1967, 651).

Many think that these are two completely different ideas. Trust-in-reliability is an expectation of performance, whereas trust-in-trustworthiness is an expectation that something with agency will consciously adhere to norms. Russell Hardin took trust-in-trustworthiness as basic, and noted that it is hard, if not impossible, to understand trust-in-reliability in its terms: "It is … implausible that we could trust enough individuals in [an] organization, each in the relevant ways, to do what they must do if the organization is to live up to our trust. For many cases, including most governmental organizations, scale factors make this epistemologically virtually impossible" (Hardin 1999, 38). Adam Seligman agreed that trust-in-trustworthiness is the only real kind of trust, while trust-in-reliability is an entirely different kettle of fish that he preferred to call "confidence" (1997, 17); others call it "reliance." One's trust of a person might be trust-in-trustworthiness (e.g. Bob trusts Doctor Alice to use her expertise to make him well), or trust-in-reliability (e.g. Bob sets his watch by Alice who always leaves for work at 8.00). In the latter case, Alice has made no commitment for Bob's benefit, and would not be blameworthy if she inconvenienced him by not leaving on time. A trustworthy agent has free will, whereas a reliable object either does not, or its will is not relevant. We react differently to betrayals and letdowns. We invest moral feeling into trust-in-trustworthiness, and it is more fulfilling psychologically, socially and politically.

Nevertheless, I prefer to follow Dr Johnson, and treat the word "trust" as ambiguously covering them both, while separating reliability and trustworthiness. Constant across trust-in-reliability and trust-in-trustworthiness is that a person, thing or system is expected to perform in particular ways, ideally to our benefit, that we are led to this belief by signals from or about the object, and that we are entitled to act as if that performance will happen.

Trust and reliance/confidence have much in common, and at the limit can be hard to tell apart (as accepted by Seligman, although for his response, see 1997, 19–20). Many technologies have social affordances and are deeply embedded in human-to-human activities, so that the dividing line between trustworthy people and reliable systems is not as clear as one might initially think (Lankton et al. 2015). Our attitudes toward cutesy humanoid robots (Marchesi et al. 2019) and pets (Slovenko 1983) are quite nuanced, and taking the *intentional stance* toward another (Dennett 1987) is not a simple matter of anthropomorphising it. The relation of many people to the natural environment borders on the intentional; sailors may treat the sea in this way, "reading" the tides, winds, treacherous currents, interacting swells, undersea phosphorescence, and the behaviour of birds and fish, as well as using memory and narratives (Strongman 2008; George 2012). In each case, the "confidence" we have in robot, pet, environment is intermediate between paradigmatic trust-in-reliability and paradigmatic trust-in-trustworthiness, part of a cluster of related attitudes.

Anthony Giddens argued that trust-in-reliability is derivative from trust-in-trustworthiness, i.e. we learn to do the latter, and expand the concept to encompass the former (1990, 97). Philip Pettit argued the other way round, that trust-in-reliability of persons is the general phenomenon of which trust-in-trustworthiness is a special case (2004, 108). Though these are in tension, I don't see that they are inconsistent, but rather intertwined. We grow up trusting certain adults, usually parents, in a pretty hard-wired way. This is likely to be part of the process by which we grow to trust the environment as a stable context for our behaviour (Giddens 1990, 94).

Important though trust-in-reliability is for negotiating the complexities of digital modernity, space prevents a deep discussion, and I will focus for the rest of Part I on trust-in-trustworthiness. Trust-in-trustworthiness demands complex and strategic thinking: Bob's trust is based not only on his expectations of the likely behaviour of Alice, but also on his expectations of her expectations of him. Hence an investigation into trust-in-trustworthiness will be more informative about our social and political relations.

Trust – Explicit and Inferred

In a situation where trust and trustworthiness are, as it were, on the surface, Alice and Bob have expectations, and can reason about them. However, the trust relationship may not be evident to the two parties, perhaps because the interaction is a familiar practice. It is not present to their consciousness, but may be inferred by outsiders analysing the practice.

The classic example has already been covered, bus travel as a form of life. Bob must trust the other passengers not to attack him, but this rarely reaches his consciousness. Normally one just hops on the bus without evidence at hopping-on time that the other passengers are trustworthy. The outside analyst, however, may be unable to make sense of the rationality of the practice in the absence of an inference that Bob subconsciously trusts his fellow passengers.

Another example is the deep and lasting bond between a baby and its mother. This is no doubt hard wired in the brain, and the infant can have no consciousness of trusting. Yet psychologists reason about the role that trust plays in child development – for instance, attachment theory suggests that children securely bound to their parents develop an internal model of others' intentions that supports trust in later life, whereas less secure children struggle in this respect

(Holmes 2014). This is evident neither to child nor parent, but there are methods in developmental psychology to measure it (Kerns et al. 1996).

As a third example, poker players may be conscious only of their mistrust of each other's bluffs and stratagems, though again it is hard to make sense of their game without the inference that they trust each other at least to follow the rules, for example playing the cards they are randomly dealt.

Inferred-trust and conscious-trust may at times contradict each other. "I trust you," Bob says soothingly to Alice, while never placing trust in her, even when it would be advantageous to do so. Should she believe his words, or his actions? Is she right to be offended?

Trusting and Placing Trust

A very important ambiguity, which often leads to misunderstanding, is between trust *as a graded attitude toward or evaluation of a would-be trustee along a scale* (for instance, low trust, neutral, high trust), versus trust *as a binary incorporating a threshold* – Bob trusts Alice, he does not trust Adele. The first is a general view of the would-be trustee, while the second is a judgment about her in a context. I think we can connect graded-trust and threshold-trust.

They also spin off a related ambiguity between *evaluative trust* and a more *action-oriented* sense. This ambiguity is between *trusting* (*believing* in the trustworthiness of another) and *placing trust* (*acting* on that belief, or, more widely, acting as if that belief is true). One's trust may be general, but placing trust only happens in a specific context. Placing trust covers any action warranted by trust, even in low or zero risk contexts, and so would include, for instance, recommending someone to a friend, or placing trust in a politician by voting for her. However, this statement masks a lot of complexity which will be unpicked in Chapter 4. Trusting is conceptually prior to placing trust, but trust placed is the canonical evidence for the beliefs associated with trusting.

It is common to refer to threshold-trust as an evaluation of the would-be trustee's trustworthiness in a context, yet graded-trust often features in our discourse, and the distinction between them is rarely emphasised. We can illustrate it with commercial lending. A moneylender is prepared to lend money to everyone except for a few recalcitrant individuals. From the binary perspective of threshold-trust, it appears that he trusts everyone, because they exceed the threshold at which he will trust them with a loan. However, this is a highly misleading description. Actually, the lender does due diligence on those he lends to, and charges interest at a rate that varies depending on his assessment of the borrower's risk of default. If his graded-trust in a borrower is low, then she will have to pay a high rate, and conversely high graded-trust in her means lower interest. The graded-trust function is reflected in the variable interest rates charged, and the lender applies the rate that means the risk of default (borrower's untrustworthiness), which acts as a discount on the expected return, is balanced by the expected reward. It is extremely hard to explain the sliding scale of interest rates on the binary model of threshold-trust.

Even if it is more commonly expressed, threshold-trust is best seen as a derivative of graded-trust. On this reading, graded-trust is a kind of function covering a range of contexts. When we fill in the context of the graded-trust function, then we are able to judge threshold-trust in that context, i.e. whether the would-be trustee will be trustworthy or not in that specific circumstance. That judgment feeds into a decision about whether to take action, i.e. whether to place trust in her in that specific case. Graded-trust may be very sensitive, where the trustor has a lot of experience and knowledge of the would-be trustee, or may be largely empty, with significant detail only for the contexts in which the trustor expects to interact with the trustee.

Local and Global Trust

Yet another distinction is between local and global trust (O'Hara 2004, 75–94). *Local trust* is based on acquaintance, founded on direct interaction with or knowledge of the would-be trustee. It may also rely on testimony of trusted friends, and the reputation of the trustee. Such close interaction is effective and satisfying, though not infallible. As trustors make more judgments about others, they become less naïve and more discerning (too many bad experiences may make them cynical and distrustful). Via high-bandwidth interactions, they receive unconscious signals of untrustworthiness, such as sweaty palms or a reluctance to make eye contact. Unfortunately, they may also be prejudiced, for example by accent or skin colour.

Local trust requires a major investment in cognitive resources on the parts both of trustor and trustee. Consequently, it is characteristic of premodern and smaller communities, where interactions can realistically be face-to-face, or based on information shared across a trusted network of family and friends. Such networks may be embedded within larger societies, such as villages, associations or tribes, or they may be the whole community. When something really matters, local trust is more comforting – we consult the doctor with whom we get on best, we ask for neighbours' recommendations of tradespeople, we cast an eye over our children's sweethearts, we interview our employees. It has sometimes been called 'thick' trust, based on intensive contact, familiarity and strong ties with others. It brings the possibility of local and meaningful sanctions, disseminating opprobrium through a small world of mutual friends and contacts.

But these cognitively-expensive methods can't sustain all the interactions of a complex and dynamic society. Decreased familiarity means that trust is all the more essential (Luhmann 2017, 25), but there are too many strangers for face-to-face interrogation to scale. Consequently, greater social complexity is facilitated by *global trust*, or 'thin' trust, maintained by methods, mechanisms and institutions that exploit economies of scale. Governments and banks may verify someone's identity; banks may also certify creditworthiness. Professional associations, guilds and educational institutions may certify skill or expertise ata trade or occupation, and may formulate codes of conduct to fix the ethical foundations of their practice. Regulators define minimum standards for goods and services. Celebrity endorsements give positive accounts. The media publicise poor performance.

Global systems are based on intermediaries between trustor and potential trustee, who help a range of trustors to make rational, self-interested judgments. The intermediary may be a person or institution, but could also be a piece of technology (such as an Internet platform). It performs tasks that support trust judgment, such as monitoring trustees' behaviour, administering systems of rewards and sanctions, certifying trustworthiness, defining the types of task that the trustee might be expected to carry out, or simply providing a context of interaction that is meaningful for both parties. One useful summary divides global mechanisms in the economic sphere into five types: market solutions (commercial information providers and certifiers); self-regulation by informal communities and networks; self-regulation by formal associations and hierarchies; state and regulatory solutions; and governance hybrids where states and civil society interact with markets and international agencies (van Waarden 2012).

Global trust is a many-to-many relationship, bringing with it both economies of scale and systemic risks. If Alice, certified as trustworthy by Carlotta, an intermediary, proves to be untrustworthy to Bob, then not only Bob but all the people Alice has let down may withdraw their trust from not only her but everyone else certified by Carlotta (i.e. they lose trust in Carlotta's recommendations). Worse, if news gets out, then *everyone* who relied on Carlotta might withdraw *all* their trust from *everyone* certified by her (i.e. trust in *Carlotta* vanishes, thanks to the untrustworthiness of *Alice*). The effect on social trust of even a small number of systemic crises may be dramatic, and there may not be alternative mechanisms to repair it. In short, you can't get rid of the need for trust by using institutions or intermediaries, for the obvious reason that you have to trust them. All you are doing is shifting from local trust to global.

Regularity and Morality

Finally, we can contrast *trust as ethically-charged*, and *trust as expectation of norm-following*. Trust thrives on regularity and normality in behaviour, yet social norms vary in terms of their ethical salience. Is trustworthy Alice good, by virtue of being trustworthy, or simply conventional? "What is expected of you is, morally, that you be trustworthy, and, socially, that you abide by the relevant norms. Is this a distinction without ultimate difference? It would be, if the demands of morality were finally those of local norms or, conversely, if local norms involved obligations not accounted for by their being sanctioned by local practice" (Hollis 1998, 11). Trustworthiness is often seen as a moral attribute, but also involves following conventions, and the distinction between conventions and moral rules may be none too clear. Many liberal thinkers, Martin Hollis included, see local norms as sources of conflict and injustice, but that perception may lead to a brand of cosmopolitanism that prompts us to misunderstand or even invalidate social behaviour in smaller communities (Deneen 2019, 24).

One way of considering this question is whether the would-be trustee acts only in self-interest, or according to her obligations (Hollis 1998, 13–14). I'm not keen on this as the basis of a distinction, because it is in our self-interest to live up to our obligations, and be seen to do so. Game theoretic approximations to social situations, such as the prisoner's dilemma or the ultimatum game (Skyrms 1996, 25–32), are unconvincing as predictions of actual behaviour, particularly in repeated contexts, precisely because it is indeterminate as to whether someone will draw her self-interest widely or narrowly. Some people will shaft you given half a chance, and others won't – but we knew that anyway.

We must note this ambiguity, which will, for example, affect the blameworthiness of Alice if she lets Bob down. However, the ethics of trust and trustworthiness will be important for our inquiry about their value, and so in the next section I will expand on the topic.

THE ETHICS OF TRUST AND TRUSTWORTHINESS

Trust and trustworthiness have sundry moral aspects; they can't be ironic, and they can't be utilitarian. One ought to be trustworthy, and in a Kantian vein we might say one ought to try to trust, even if they are not ultimately winning strategies. "Any moral element in trust needs to be the genuine article and not just an ingenious remedy for a myopia endemic in philosophical egoists" (Hollis 1998, 98). In practice, our decisions to trust or to be trustworthy often depend partly on calculation, partly on instinct, and partly on a generous desire to do the right thing, and our self-interest often combines prudence and generosity.

The Virtues of Trust

Many commentators do not accept that trust is necessarily praiseworthy. For Hardin, "writers often transfer to trust the moral approbation that might be thought applicable to trustworthiness … [but] trust might be fully explicable as a capability or as a product of rational expectation without any moral residue" (1996, 28). Luhmann focused on trust's role in reducing the complexity of modern life – handy, but not a moral good (2017, 27–35). Trust may also go wrong, or be associated with negative outcomes. For Baier, it "is not always a good, to be preserved. ... [T]he first thing to be checked, if our trust is to become self-conscious, is the nature of the enterprise whose workings are smoothed by merited trust" (1994f, 130–131).

Trust is often a response to uncertainty, and the epistemological aspect does bring in one type of virtue – epistemological virtue (Sosa 2007). The trustor's judgment may exhibit some of the epistemological virtues, such as conscientiousness, open-mindedness, objectivity, and so on, in which case it would be praiseworthy, or vices, such as cynicism or gullibility. There is a debate about whether virtue epistemology falls within virtue ethics (Swanton 2019); certainly Aristotle treated epistemological virtues as intellectual virtues in his ethical writing. They can clearly contribute to the quality of the trustor's life, as virtues are supposed to do.

The act of placing trust also involves some of the more traditional moral virtues. For instance, it partakes in two of the three Christian virtues: faith (confidence in a person or project) and hope (the optimistic desire for something positive). "The function of faith is not to reduce mystery to rational clarity, but to integrate the unknown and the known together into a living whole" (Merton 2007, 136). The religious view carries over into secular contexts: "Trust is not the same as faith in the reliability of a person or system, it is what derives from that faith. Trust is precisely the link between faith and confidence ... . *All* trust is in a certain sense blind trust!" (Giddens 1990, 33, his emphasis). Hope is a forward-looking idealisation, essential to planning, design, innovation, travel and human interaction (McCloskey 2006, 160–163), the belief that "As good as the world is, things are going to get even better and we can make it so" (Uslaner 2002, 2).

The alternative, refusing to have faith, is self-defeating cynicism, as in Baier's account. "If trust is seen as a variant of the suspect virtue of faith in the competence of the powers that be, then readiness to trust will be seen not just as a virtue of the weak but itself as a moral weakness, better replaced by vigilance and self-assertion, by self-reliance or by cautious, minimal, and carefully monitored trust. The psychology of adolescents, not infants, then gets glorified as the moral ideal" (Baier 1994e, 108). Self-reliance which includes such suspicion of faith inhibits trust as a valued and valuable social resource, as will be argued in detail in Chapter 10.

Additionally, placing trust implicates the classical cardinal virtues, balancing courage and prudence. When the trustor risks a lot, he is courageous (and possibly foolhardy, when courage is in excess); when he is cautious, he is prudent (and possibly timid, in deficiency). When prudence is unaccompanied, it "makes it rational to act in a trustworthy way less often than it takes to sustain a general trust" (Hollis 1998, 5; McCloskey 2010). On the other hand, human development requires trust without prudence, as in an infant's trust in its mother (Giddens 1990, 94–97; Baier 1994b, 42–43; Hursthouse 1999, 225–226). Baier's cynical characterisation of the "psychology of adolescents" is upstaged by her argument that "unforgiving rigidity and, at the other extreme, easygoing willingness to keep on forgiving are both dysfunctional weaknesses, if our goal is to maintain and repair a network of beneficial trust … composed of normally faulty human persons" (1994f, 150), which rather better describes the excesses and deficiencies of prudence in the virtue-complex of trusting.

The Virtues of Trustworthiness

It would seem clearer at first blush that trustworthiness is a trait of a virtuous individual (Lenard 2012, 33). Deficiency of trustworthiness is the vice of untrustworthiness, and its excess is the vice of inflexibility. In moderation, trustworthiness is part of a person's excellence, and the good habits thus cultivated, directed toward others, will play their part in creating and cementing communities, the vital function of virtuous behaviour. Yet this view is not widely held (Jones 2012, 78–85; Hawley 2019, 76), because trustworthiness, like trust, may enable bad things to happen. A criminal gang may be highly trustworthy to each other, but itself is antisocial.

We should emphasise that trustworthiness largely produces good results, and the occasions when it doesn't are surely in the minority. Still, it is hard to do only good, and any kind of virtuous behaviour can have ambivalent effects. Virtue is complex: "no virtue stands alone, … every virtuous act costs something in terms of other virtuous acts, … virtues are intertwined with evils, and … no theoretical system of a hierarchy of virtues is ever realizable in practice" (Shils 1997a, 52).

Space precludes deep discussion, but I count both trust and trustworthiness among the virtues, and rely on virtue ethics to counter some of the arguments about their disadvantages. The dispositions underlying virtues are deep, steady, intelligent and reliable, not simply whims or habits (although they may include habits). They are also broad, informing attitudes, emotions, reactions, plans and desires. Some have suggested that virtuous people are good only when their motivations are good (Slote 2001; Hills 2023, 760), but that excludes those who are trustworthy for money, all sorts of professionals and craftspeople – doctors, accountants, bureaucrats, plumbers, mechanics, air stewards and very many others who would not be trustworthy in their jobs unless they were paid sufficiently. They go on strike, they resign, they retire, they act as jobsworths, they finish up for the day and forget their customers' interests. Yet such people may still have a strong vocation beyond the merely transactional, even though, in the absence of the transaction, they would have no desire to do a good, or even a bad, job at all. The virtues of those working in institutional, governmental or commercial contexts need not involve high moral motivations, but may be as simple as conscientiously prioritising some goods of a practice over others, a more modest casting of virtue (Zacka 2017, 138–140; Swanton 2021).

Trust, Power and Coercion

Trust can go wrong when it shades into coercion. Bob may trust Alice because she has groomed him, or Alice may have been manipulated by Bob into behaving as he demands. Baier warned against relationships marked by power and dependency (Baier 1994e, 106–110), partly because she saw them as characteristic of Western patriarchy, hinting without actually stating outright that abuse is the norm in "the sorry sexist history of marriage" (1994e, 120–123). Yet power asymmetries are normal in all sorts of trusting relationships. Most people want to be married (and employed, and treated by doctors, and taught by teachers), so we won't follow Baier in throwing out the socialised baby with the abusive bathwater.

She was correct that abuse is to be guarded against, but we cannot draw any conclusions directly from power asymmetries. The paradigm is the infant's total, unconditional, inexpressible and hard-wired trust in its mother (Erikson 1963, 222–226; Waters & Deane 1985; Cohn 1990), but there are plenty of other examples: Bob the aspiring sportsman's trust for Alice, his Svengali-like coach; ill and housebound Bob's trust for Alice his carer. In these cases, the trustee frames the relationship and shapes the trustor's interests accordingly. Yet similar cases may be manipulative, such as the overbearing husband manipulating his wife into becoming what he sees as a trustworthy and properly deferential life-partner and support.

Coercion is a problem when prudence goes too far, and trust is taken consciously as a means to an end, rather than the social reflex it often is. An influential account called *encapsulated interest theory* describes how the trust relationship aligns the interests of rational people:

> I trust someone if I have reason to believe it will be in that person's interest to be trustworthy in the relevant way at the relevant time. My trust turns, however, not directly on the Trusted's interests per se, but on whether my own interests are *encapsulated in the interests of the Trusted*, that is, on whether the Trusted counts my interests as partly his or her own interests *just because they are my interests* … . (Hardin 2006, 19, his emphases)

In other words, Alice's ("the Trusted") trustworthiness consists in her adopting Bob's interests as her own. This makes her trustworthy, on the encapsulated interest account, assuming she pursues her own interests in a rationally self-interested way, because in pursuing them she also pursues the relevant interests of Bob.

While this no doubt describes some interactions, it is hardly plausible as a general account, partly because it minimises, or even denies, the virtuousness of the trustworthy. If trustworthy Alice is merely pursuing her own interests (which encapsulate some of Bob's), then she isn't really going out of her way or doing anything difficult in being trustworthy. In a Kantian mood, it's far less virtuous to pursue *your own* interests (even if adopted ones), than to pursue interests of others that you do not share.

More importantly, there is something manipulative about Bob's relation with Alice, as with the overbearing husband. Friedrich Hayek defined coercion in the following way:

> Coercion occurs when one man's actions are made to serve another man's will, not for his own but for the other's purpose. It is not that the coerced does not choose at all; if that were the case, we should not speak of his "acting." … Coercion implies, however, that I still choose but that my mind is made someone else's tool, because the alternatives before me have been so manipulated that the conduct that the coercer wants me to choose becomes for me the least painful one. (Hayek 1960, 133)

If we apply Hayek's definition to the case where Alice has encapsulated Bob's interests in her own, we find her the victim of Bob's coercion. Her actions serve Bob's will for his purposes. I would like to resist the conclusion that trust is a kind of coercion or moral blackmail, and it's not a good look for a theory of trust to morph so easily into a theory of coercion.

We should therefore reject encapsulated interest as a general account. In civil society, the trustee chooses whose interests she will pursue (or agrees to respect), perhaps because of interests of her own, or because she feels an obligation. If Bob is a murderer defended by Alice the lawyer, her interests surely cannot encapsulate his "just because they are his interests." Rather, as a trustworthy lawyer, she is committed to pursuing the interest of the court – that is the nature of her oath – which is that justice emerges from competent argument on all sides, including on behalf of the guilty, the feckless, the uneducated, those with psychological difficulties, foreigners and others who would struggle with its conventions. Alice's personal interests are surely better served by Bob's being put in prison, thereby rendering her a little bit safer. She puts these aside (as well as personal reluctance or risk of reputational damage) to ensure that Bob has a proper defence.

However, we shouldn't go too far in the opposite direction. Luhmann stated that it *cannot* be trust if the trustee "will be likely to toe the line on his own account" (2017, 47), i.e. Bob can't trust Alice if she is likely to do what he wants her to do anyway. This is the opposite of Hardin's view, but is still incorrect.

Bob still needs to trust Alice because in general he cannot know what her interests are. At a minimum, he can never be 100% sure that she hasn't simply been bribed to act against his interests, and even against her own. If their interests coincide, he may not *need* to trust her (Luhmann's point), but he cannot know that for sure. There is still uncertainty which must be bridged by trust.

Virtue and Rightness

Many of the problems highlighted in this section stem from a desire for virtue theory, as a moral theory, to tell us the right thing to do, but there is no reason to expect this. After all, identifying the right thing is tricky on any theory, which is why moral philosophy is difficult. Even a flexible account can require complicated and counterintuitive caveats to deal with tragic dilemmas (Hursthouse 1999, 44–51). Virtue theory tells us more useful things about the excellences of certain practices. In particular, achieving excellence usually demands further virtues, so that one virtue leads to others via the things we try to do well in daily life. The purely antisocial trustworthiness of the criminal gang to each other is an isolated pattern of behaviour, and therefore harder to maintain. That said, there are certainly vocations where the excellences can be manifested to ambivalent effect, including finance and computer hacking (Himanen 2001).

Modern ethical discourse often focuses on the decontextualised actions of individuals, and moral rules aspire to generality and widespread applicability to choices and actions. Surely, to the contrary, most *behaviour* takes place within particular institutions, practices, families, companies, jurisdictions, etc., and many important projects involve the construction and sustaining of idiosyncratic forms of community that enable shared achievements and common goods. Our ethical selves develop over time in such *milieux*, through our choices and the building (or neglect) of our character. A person need not even see herself as virtuous when acting virtuously – the virtue may be delivered by habit, intuition, practical rules of thumb, or social norms (Burke 1999a, 180–183; Hursthouse 1999, 127–128). We do our ethics by living within the roles we occupy, and the standards we are set. If we don't take those roles seriously, then we may unwittingly allow our virtue to decay into excess or deficiency. For instance, a bureaucrat's disinterest may become indifference, predictability become mindlessness, justice become rigidity, caring become ingenuousness and naïvety (Zacka 2017, 66–110).

Most moral philosophers either wish to reconnect virtue and right action, or alternatively to reject virtue because it cannot be securely connected to rightness. But the minority view of Elizabeth Anscombe (1958) that we can and should reject *rightness* to the extent that it is *not* connected to virtue has, it seems to me, a great deal of merit. Not least, we can preserve the straightforward intuition that trustworthiness and trusting are virtues.

TRUST AND MODERNITY

The final preliminary argument that I wish to make is that trust and trustworthiness are universal or near-universal aspects of human nature. Again, this may seem obvious, but it has been argued that they emerged only in modernity, and are essentially products of complex social arrangements. Premoderns, "conditioned by the mutuality of systemically reinforcing role expectations" (Seligman 1997, 52), are in thrall to the demands of the roles they occupy. They can therefore be reliable or unreliable, but cannot be the objects of *trust* because they are not afforded agency by the systems they engage in. On this view, the objects of trust are *exclusively* modern individuals, whose market exchanges governed by self-interest are not restrained by solidarity (Seligman 1997, 89). "The 'trust' that is so often seen to bind members of tribal, peasant or other types of premodern societies is not trust at all but confidence in a very particular mode of social organization based on ascriptive categories (the same would hold for contemporary Japan)" (Seligman 1997, 37; Reemtsma 2012, 10–12, for a variant on this position). Seligman accepted the unpalatable corollary that trust "is far less pervasive than we would like to think" (1997, 28, 70).

Premodern Trust

Seligman's curious downgrading of modern Japan is the first hint that this can't be correct. Certainly trustworthiness plays an important part in the proper and effective functioning of the market, yet it seems implausible to suggest that this is its only relevance and first historical appearance. Granted, the modern individual has many roles and is distinct from them, but was this not true in premodern times? The essence of tragedy is the juggling of roles. Lear drops the role of king, and struggles to maintain either the role or authority of a father; Antigone buries her brother, preferring the role of sister to that of obedient subject. Greater expectations may have hung on the occupancy of a role, and it may have taken Machiavelli to point it out, but premoderns would see betrayal and unmet commitments all the time.

Knights, governed by the ideal of chivalry, "were supposed, in theory, to serve as defenders of the Faith, upholders of justice, champions of the oppressed. In practice, they were themselves the oppressors, and by the 14th century, the violence and lawlessness of men of the sword had become a major agency of disorder" (Tuchman 2017, xxvi). In medieval times, before the invention of the enclosed confessional, women's confessions were supposed to be heard in public in case the priest tried to abuse his position, and there are documented cases of priests attempting to seduce the confessor in private confessions (Orme 2021, 269). In a 13th century fair in England, "legal coercion was deployed consistently with informal judgments of reputation. One known to be untrustworthy … might be unable to find pledges, or others willing to risk damnation by swearing an oath in his favor" (Sachs 2013, 27). Someone's reputation must be distinct from their occupancy of a role, as it is an evaluation of their record in role-occupancy. In premodern times, knights, priests and merchants alike often behaved in unknightly, unpriestly and unmerchantly ways. There was always a private person behind the mask, if less valorised in premodernity than in modernity (O'Hara 2023, 30–31). Trustworthiness may well rest on the agency of the person (cf. Seligman 1997, 63), but agency is not a modern invention.

The Blurring of Trust and Trustworthiness in Modernity

I, like many others (e.g. Hardin 1996), insist on the clear distinction between trust and trustworthiness. Unfortunately, when one pursues the idea of trust as a product of modernity (Seligman 1997), and Enlightened rationality (Hollis 1998), the distance between them seems to melt away. Hollis imagined a series of abstract thought experiments in which most of the facts are known to both trustor and trustee, so the *trustee's* reasoning about whether or not to be trustworthy is open to the *trustor's* analysis. Furthermore, the *trustee* can walk through the *trustor's* analysis, because she can anticipate his anticipation of her own initial reasoning. They go together in lockstep, each anticipating the other's actions and betrayals. Trust and trustworthiness become two sides of the same coin, open to both trustor's and trustee's reasoning.

Seligman was also impressed by agents' capacity for stepping out of their roles altogether, and thereby creating a space for betrayal (1997, 62–65). But both trustor and trustee have that capacity, so that if it is the main marker of a *trust* relation (as opposed to one of *reliance*), it cannot distinguish very well between trust and trustworthiness. Each becomes a type of *attachment* to a role nexus or what I called at the beginning of this chapter a form of life. On this view, they are *attitudes to* forms of life and the roles they specify, viz., determination to stay in-role (which invites each to reason about whether it is rational for the other to continue to occupy the role, as in Hollis' account). I would rather say that trust and trustworthiness are *embedded in* forms of life, making sense within them, and being defined by them (therefore opening up the possibility that they are quite distinct).

There is a deeper lesson here. If trust is a peculiarly modern and rational phenomenon, associated in particular with reason, individualism, capitalism, scepticism about authority and so on, then it seems that the particular type of value it implements is bound up with the main aspects of modernity. On the other hand, if trust, as I would argue, is more or less a constant component of all functional civilisations (a *sine qua non* of civilisation itself), then it is hardly a *creation* of liberal modernity, but rather *something upon which liberal modernity rests*. This will turn out to be a key point in Parts II and III below.

3. Trustworthiness

This chapter will describe the properties and markers of an agent's trustworthiness. Trustworthiness is concerned at its core with the expectations of others, which may be disappointed accidentally, deliberately or negligently. It is a more complex attribute of (usually) people than reliability is of (usually) objects or systems. It's a nice philosophical point whether non-humans (e.g. AI) can be called trustworthy, as opposed to reliable, but I won't go into it here (Pettit 2004; Hawley 2014, 2; Al 2023). I will occasionally refer to a system or an organisation as trustworthy; if that offends you, then please adjust the vocabulary accordingly.

Trustworthiness often has a moral overlay around its expectational core (Hardin 1996), concerned with protecting the moral order and the intuitive fairness encoded into social conventions and norms. Following such norms means respecting one's society and the members of one's community. From that perspective, trustworthiness is a bound upon liberty and self-expression. The moral aspects of trustworthiness have been discussed in Chapter 2, and will be revisited many times; this chapter will be concerned with the basic processes of expectation management that comprise trustworthiness.

THE PARAMETERS OF TRUSTWORTHINESS

My account is similar to Katherine Hawley's (2019), that if Alice is trustworthy, she will in general keep her commitments to others. She will do what she says she will do. However, this blithe description leaves much unsaid, and there are many variations on this simple theme. In this section I shall work through its various parameters.

The Agent

A person (or thing) that is trustworthy must have agency. If Alice is trustworthy, she will either do what is claimed for her, or there will be a reason for her failure. Not all agents are trustworthy. Circumstances may prevent Alice from doing what is claimed for her, or she may use her free will to refuse.

Alice's trustworthiness may be a property of her character, exhibiting the virtue of trustworthiness. Alternatively she may occupy a role (e.g. a bank manager, a policewoman) in which she is trustworthy, and her trustworthiness is shown by her occupancy of the role in good faith. Her personal trustworthiness would be exhibited by her occupying the role in a trustworthy way; the role itself would determine what her commitments are. If she leaves the role, e.g. of bank manager, she ceases to be a trustworthy bank manager, and her replacement Adele would become the trustworthy manager behind the desk.

Roles usually sit within a nexus of relations in some kind of system. If the banking system is reliable, that lends support to claims of trustworthiness for those with roles within it, because a reliable banking system delivers the financial outcomes that it promises, for which the role-occupants need to play their parts. As another example, if scientific research and peer review make up a reliable system, then scientists might be said to inherit trustworthiness from it, and even pieces of research can be reliable, as long as they were derived with respect to the conventions and constraints of the wider system. Many claims of trustworthiness depend upon a background of system reliability.

Relatedly, we may also see groups or classes of agent as trustworthy. Perhaps journalists, politicians, newsreaders, priests, doctors or soldiers might be counted as trustworthy (or not). If Alice is, say, a journalist, then she may be judged *prima facie* as trustworthy as her profession. Further evidence may enable more personalised judgment later on.

Where Alice's trustworthiness is linked to her character, then this is likelier to be perceived most clearly by those with direct acquaintance of her. Where it depends on her role, or on a background system, then direct acquaintance is less likely to be needed.

The Commitments

A trustworthy person is trustworthy in some respect or other. One might be a trustworthy mechanic, but not a trustworthy brain surgeon. I follow Hawley in using the term "commitment" to refer to whatever it is one is trustworthy about.

In my original version of this theory (O'Hara 2012), I used the more neutral term "representation," which I prefer to "commitment." A commitment implies specificity of what is to be done and for whom. This made it harder for Hawley to distinguish trusting someone and placing trust in them (Hawley 2019, 9). Alice's being a trustworthy plumber, for example, doesn't mean she has any plumbing commitments to anybody – she may be on holiday, or ill, or working towards a career change. Her trustworthiness is a subjunctive claim about how she *would* respond *were* she to have specific commitments. It might be described as a disposition, although that emphasises a behavioural dimension that we may want to play down.

A focus on *specific* commitments draws attention from the *range* of desirable things that a trustworthy person might do. Someone may demonstrate her trustworthiness by using her discretion, knowing when to step away from or transcend specific commitments and to act so as to benefit a wider or more deserving set of people (Kirton 2020). Trustworthiness requires judgment: if Alice "remembers" someone else's birthday by giving her bank a standing order to send flowers on the correct day and never thinks about it again, we wouldn't see this necessarily as trustworthiness on her part (Baier 1994f, 136). This is what Alison Hills called taking *responsibility* for a task, arguing that it is different from being committed (2023, 750–751). Commitment must be taken widely and generously to cover cases where, as Hills put it, responsibility outruns reliability.

Another aspect of trustworthiness, perhaps an essential one, is honesty (Rotenberg 2010). A trustworthy person will not mislead others, and will not be ironic, "knowing," satirical or legalistic (Reemtsma 2012, 22). Is this a *commitment* to others or a character trait? We could argue that asserting involves making a commitment for the truth of what is asserted, taking responsibility for its truth and simultaneously discharging the commitment by the act of assertion (Williamson 2000, 268–269). Again, we must interpret "commitment" generously to ensure that honesty is covered by it.

Commitments may be held with others (Searle 2010, 42–60; Gilbert 2014a), which affects the trustworthiness of associations and the individuals within them (Gilbert 2014b). For instance, if two colleagues make a joint commitment to a client, it isn't reducible to two separate commitments, as it also includes commitments to each other. Their trustworthiness could easily be conflicted – for instance, if one reneges, will the other inform on her to the client, take a share of the blame, or try to cover up?

Hawley recognised some of these issues, leading her to introduce odd augmentations like "meta-commitments" (Hawley 2019, 73–79), and she struggled with the idea that there can be non-voluntary commitments (2019, 98). So "commitments" is a problematic term, but Hawley's account is deservedly well known, so I will adopt her term with reservations. It implies better than my term "representation" the nature of trustworthiness as a virtue (Hills 2023; although Hawley herself thought not – 2019, 76, 84), and has a natural extension to mistrust and untrustworthiness (Hardin 2006; Hawley 2019, 16–20). "Commitment" must, however, be given a broad interpretation: "commitments can be implicit or explicit, weighty or trivial, conferred by roles and external circumstances, default or acquired, welcome or unwelcome" (Hawley 2019, 10), and additionally imposed by mutual expectation and social convention.

Finally, it is worth flagging the distinction between individual commitments and the general quality of commitment. Given that we carry out our commitments, we can be *more or less committed*. If Alice has committed to a task

– say, caring for an elderly and ailing relative – she might carry it out *with commitment*. She rolls up her sleeves and does the work herself, cheerfully and to the best of her ability, showing a strong commitment to her relative's comfort, health and emotional well-being. However, her commitment would also be discharged if she merely paid a firm for nursing care of a sufficient standard. She might argue that the cost in money is greater than her investment in time would be, but there is a sense that paying for help shows *less* commitment to the task than doing it herself (though cf. Uslaner 2002, 136).

The Stance Toward the Commitments

Suppose Alice to be trustworthy and to have some commitment. In that case, it follows that she *intends* to carry it out, even if she would rather be doing something else. She is willing and ready to do it.

Secondly, she must be *able* to carry it out. If it is not within her gift, then she cannot guarantee to do it, and may be derailed in a number of ways. Perhaps she overestimates her powers, or the context changes, or maybe she is prevented. Maybe she has taken on too many other commitments, and has a diary clash. If her commitments come from a role she occupies, it may be that the role imposes too much on her or demands different priorities.

Third, she must be *motivated* to do it. Something more important may come along, or she may be waylaid by a higher duty, such as rescuing a struggling child from a duckpond. Maybe it is as simple as her not being paid enough to do it. Maybe the commitment was made so long ago that she has discounted the incentives to the point where they have lost their significance. Maybe the person to whom she made the commitment has let her down in the interim.

Note that not all incentives or motives to carry out one's commitments are financial or otherwise self-serving. The fact that she has made the commitment *in and of itself* provides an important motive for carrying it out. A commitment includes within it an incentive to keep it. That's what makes it a commitment as opposed to an airy assurance: one of the differences between making a commitment and pretending to make one is that in the latter case one provides oneself with no motive for carrying it out.

If Alice is willing, able and motivated to carry out the commitment, then she will, all things being equal, carry it through. If she fails, then it won't be her fault. Even trustworthy people fail to carry out their commitments sometimes, but – if they behaved in a trustworthy way – they are not responsible when things go wrong. Either the blame should fall elsewhere, or it couldn't be helped. A standard vocabulary of excuses has evolved over the years to deflect blame along these lines, including "events beyond our control," "unforeseeable circumstances," "teething troubles," "staff shortages" and "the computer is down" (Hood 2011, 52).

The Content of the Commitments

A trustworthy person is committed to *something*. This may be specific, such as to decorate a particular room by the end of the week, or more general.

One weakness of the term "commitment" is the implication of specificity, because a trustworthy person's commitment may be broader: to be a good friend, to uphold the law or to be a reliable financial advisor. It might be psychobabble, such as "I will always be there for you." In such cases, the commitment will unfold and evolve. Sometimes it will be obvious what Alice should do; if she is a good friend to Bob, she should help him through a bereavement, lend him small quantities of money, put herself out for his benefit. Sometimes it will be less obvious; if she knows that his wife has been unfaithful, she might be uncertain whether she serves his interests more by telling him or concealing it, a moral conundrum that has faced many a trustworthy friend. Such commitments are open-ended (cf. Walker 1995).

On the other hand, the commitment has limits that shouldn't be exceeded without explicit agreement. "The babysitter who decides that the nursery would be improved if painted purple and sets to work to transform it, will have acted, as a babysitter, in an untrustworthy way, however great his goodwill" (Baier 1994e, 101).

As many commitments are implicit and as all demand interpretation, they have a social aspect. The content of Alice's commitment is not entirely in her hands. If she is a trustworthy friend, advisor, grandparent, role model, patriot, teacher or community leader, then she will have her own view of what they entail, but hers is not the final word. The nature of her commitment will depend at least partly, sometimes wholly, on its social construction, embedded within what John Kekes has called a *moral tradition* (Kekes 1998, 111–117). She will carry out the commitment in her own way, using her judgment and interpretation, but she also has to take into account those aspects that are socially determined.

The Origin of the Commitments

In many cases, Alice takes on commitments by her own decision voluntarily, but she may not be fully in charge of what she takes on, for example when she is the occupant of a role. Her boss, Carlotta, may make commitments for her, or at least shape them. It may be Carlotta who receives a request for help, and then assigns her employee Alice to the job. Depending how she communicates to Alice, the responsibility for a failure of the latter's willingness, ability or motivation in such a case is at least partly down to Carlotta, while Alice's responsibility is correspondingly reduced. If she isn't paid enough to do a good job, she may not be motivated to carry it through. If she is undertrained, she may not be able to do it.

Carlotta must have the *standing* to make *legitimate* commitments for Alice. There must be a relationship between them which allows her to commit Alice in a binding way – it may be that she is Alice's boss, her wife, her guru, or whatever. Even when she does have sufficient standing, Carlotta can't make all Alice's commitments for her. If she is the boss, she can commit Alice to carrying out the business of the company in office hours, but she can't commit her to do things where she has no authority, such as vote Labour, or run the London Marathon.

If Dinesh has no such standing, he cannot commit Alice to anything. Even if they have some sort of a relationship, say, drinking buddies in the same pub, Dinesh cannot commit her to playing darts in the team on Thursday evening.

Roles cannot determine exactly what people will do, and occupants have their own interpretations of what their roles require, selectively emphasising some aspects at the expense of others. A trustee of a fund is required to manage it for the beneficiaries, but how she does this is her own decision, as long as she is not negligent (Penner 2016). Most people occupy many roles, and it may be that they interfere with or influence each other, with the potential for inconsistency or conflict of interests (Seligman 1997, 29). The division of labour requires trustworthiness of all the specialised participants, and consistency between their interpretations, to succeed (Baier 1994e, 96).

Practical circumstances can affect which commitments we take on, and also our ability to carry them out (Hawley 2019, 99–105). A poor person may find it difficult to turn down an opportunity, while someone in a dense culture of reciprocity may find herself practically required to return favours. The Bedouin Balga tribe of Jordan imposes commitments on all its members to provide abundant hospitality for strangers (Shryock 2008). Complex social systems have evolved in communities via an ethic of vendetta to pursue local versions of justice (e.g. Byock 2003). Someone who finds herself a trustee in the middle of such a network of commitments needs insight into those she has taken on or inherited, and will struggle to be trustworthy in its absence (Hawley 2019, 114–116).

Finally, some commitments may not be *taken on* at all, and may not be explicit. A parent has commitments to her children, and *vice versa*, that can't easily be rejected, or even listed (*pace* Harlan Howard's country music classic "No Charge"). A citizen has commitments to her country. A member of a community has commitments to her fellow members and to the health of the community itself. A practitioner is committed to the sustainability of the practice, and to preserving its reputation, because the practice couldn't flourish otherwise. A professional sportswoman is nowadays taken to be committed to being a role model to fans. Some commitments follow from others; for instance, joining a company as an employee may result in commitments that couldn't have been anticipated at the point of signing the contract.

The Context of Application

The commitments come hedged by *context* (which, as I mean it, has analogies to Luhmann's *systems*, although not necessarily as abstract or indeed systematic – Luhmann 2017). A trustworthy person like Alice is willing, able and motivated to carry out commitments *within a reasonable framework*. Where the commitments are from her employment, her trustworthiness will only be guaranteed, as it were, in office hours, as she also has a private life into which her public role's commitments do not penetrate. Where the commitments are family- or friend-oriented, such limits do not apply to her intimate circle. The context of some commitments is society's wider moral tradition (Kekes 1998, 111–117). Some commitments may be contextualised by reciprocal arrangements – Bob will help Alice while she reciprocally helps him. If he lets her down, then she may feel her commitments annulled.

The content of Alice's commitments varies across contexts. Given that she is committed to do something, then she must interpret her commitments *in context* to decide exactly what the situation requires, and therefore how to act. In an emergency, for example, she may be justified in taking more risks and cutting more corners than would be ideal if she had time and space to plan and carry out the task thoroughly. The form that her trustworthiness takes, the way she carries out her commitments, will depend on her interpretation of the requirements of the context. This hermeneutic condition cannot be excised from the situation.

One particular type of context for Alice's commitments is an *institution*, social infrastructure that constrains or frames behaviour with roles, norms, incentives, rules, conventions, cultural ethos and ideals. Examples include families, laws, churches, government agencies, companies, clubs, armies, banks, hospitals and even etiquette. If Alice is in the domain of an institution, it might provide her with incentives to behave in particular ways, or alternatively she may use its resources in her actions. It may provide her with a set of ideals which she aspires to live up to, or furnish her with a template of how to act.

Another important context is a *practice*. It will have its own virtues, behaviours and habits which distinguish it from other practices, enhancing the mutuality of the community of practitioners, and helping mark out what trustworthy practice consists in (Turner 1994, 7–8; MacIntyre 2007, 181–203; Lutz 2012, 118–122). Explaining practice and habit across cultures is particularly taxing (Turner 1994, 20–24), and even if explicit rules exist, they have to be interpreted. That is not to say that we have to go through a process of interpretation every time we follow rules: "there is a way of grasping a rule which is *not* an *interpretation*, but which is exhibited in what we call 'obeying the rule' and 'going against it' in actual cases" (Wittgenstein 1958, §201, his emphasis). We need to look at "training and behaviour, use, custom, institution, practice, technique, agreement. The following of a rule is a custom, an institution, embedded in the agreements, in the correspondences of behaviour within a society" (Nyíri 1982, 58; cf. Turner 1994, 72).

In many cases, the limits of the commitment are assumed, vague, implicit or understood, in which case the exact way it is implemented, or whether it should be neglected, may depend on instinct and mastery of the context. For instance, Alice and Bob are in the pub, buying each other drinks in turn. It is now Alice's round, but – awkward! – a group of Bob's friends enters. Does Alice buy the bigger round for what, for her, is a group of strangers? Or does the round revert to Bob? Will they split it between them? Let the friends work out their own separate round? Let the friends treat Alice? There is nothing in the commitment to help them decide, and they will work it out on the fly.

Finally, when the context Alice finds herself in is outside the set of contexts in which her commitments are valid, she is no longer committed. An air hostess does not have to help passengers outside the aeroplane, while a parent's commitment to feed her children ceases when they reach responsible adulthood.

The Audience

Trustworthiness is relative to a commitment, which must be made *to* somebody, whom I will refer to as the *audience*.

The audience may be a general population. When Alice drives her car, her commitments to follow the rules of the road, and to avoid accidents, are to everyone. The audience is universal, targeted generally.

The commitments may also target a particular audience of beneficiaries. Alice commits to being a good parent only to her own children, not to others'. Alice the accountant commits to working in the interests of her paying clients, and won't guarantee to use her skills for others *pro bono*. Alice the military officer has commitments to her service, comrades and country, but is not similarly bound to other armies. Alice the Parliamentary candidate makes commitments at the hustings, some to a subset of voters (e.g. the working class), and some to everyone (e.g. to reduce inflation). She is a good friend only to her friends.

A trustee usually has goodwill toward her audience, and this applies especially in close, intimate relationships (Jones 1996, 8). While these are often the most rewarding, they are not typical of trust. Many trusting interactions are transactional, and Alice may see Bob unsentimentally as an income stream. That doesn't stop her including him in her audience.

The commitments typically (if not always) favour the interests of the audience. The audience expects to benefit from Alice's trustworthiness, while those excluded should not. What is beneficial may be disputed between Alice and her audience. Alice the trustworthy schoolmistress may see Bob's long-term benefit in being made to study hard and learn his times tables, while Bob's more short-term view may stretch only as far as the cricket pitch. Alice the addiction therapist may have a very different take from Bob the addict on where his interests lie.

The nature of Alice's practice can radically change the discharge of her commitments depending on what she takes as her audience. Suppose she is an accountant who makes commitments to her clients to measure, audit and improve their financial affairs. If she follows generally accepted accounting principles, her audience only covers those whose affairs are legal. However, she may stretch her audience to include anyone who pays her fees, no questions asked. There are different shades of shadiness, from questionable-but-just-about-legal to downright crooked. Each type of accountant makes commitments to their clients, but differ as to "how far they will go." Crooked Bob will probably have to pay a premium fee for shyster Alice's services.

As an accountant doing Bob's books to his benefit, Alice is trustworthy *to him*. However, her intended audience may, and ideally should, include her profession, whose good name she is committed to preserving, in which case part of her commitment is to ensure that the financial system as a whole functions fairly and legally. Alice would be trustworthy for Bob *within that context*, so if she uncovers crooked dealings, her commitments to the wider audience would eclipse those to Bob. Crooked Bob would prefer her to audit his financial affairs in such a way as to conceal or exonerate him of his crimes, but *he* cannot trust honest Alice for such services, even though in general *she* is clearly trustworthy (Hawley 2019, 18–19). Whether a practice or profession operates for the public good or exclusively for the good of its clients will depend on a close examination of its norms and the ways these are interpreted by practitioners (Schoeman 1992, 82–83).

UNTRUSTWORTHINESS AND THE ABSENCE OF TRUSTWORTHINESS

The flip side of trustworthiness is *untrustworthiness*. This is not the mere absence of trustworthiness. It is false to say that Donald Trump is an untrustworthy car mechanic, because he has never committed to being one. Commitments have to be taken on and accepted. In the Patricia Highsmith novel *Strangers on a Train*, Guy declines a commitment to kill Bruno's father; he is not a trustworthy assassin, but neither is he an untrustworthy one.

If Alice has not taken on the commitments involved in X-ing, then we must deny that she is a trustworthy X-er, but it would be misleading, and a slander upon her, to say she is an untrustworthy one. To be untrustworthy, she must take on commitments despite being either unwilling, unable or unmotivated to carry them out. Untrustworthiness is relative to the audience of the commitments.

Untrustworthiness is usually a moral failing, the vice opposite to trustworthiness' virtue. The untrustworthy may be negligent, or actively wish to mislead for personal advantage. In Dante's *Inferno*, the 9th circle of hell is reserved for the especially evil who deceived those to whom they were bound in trust: relatives, associates, fellow citizens, guests, masters and benefactors. In the absolute centre is Lucifer, bound by ice, eternally chewing on the three worst traitors in history, Brutus, Cassius and, of course, Judas.

It may be the case that one chooses untrustworthiness to an audience for a higher cause, as for example in E.M. Forster's aspiration "If I had to choose between betraying my country and betraying my friend I hope I should have the guts to betray my country" (Forster 1972, 66). However, such pleading may be a rationalisation of a more selfish frame of mind. Some convicted criminals cite important commitments to an audience that required their untrustworthiness to the general public, such as a need to maintain a lifestyle for the benefit of family or children, protecting friends or even giving ill-gotten gains to charity (Ameer & Othman 2022). An untrustworthy person "will hardly be disposed to look at himself from the perspective of distrust and seek the cause of it in himself" (Luhmann 2017, 82).

Untrustworthiness is not always morally charged; it may be down to misapprehension or incompetence. Alice might take on commitments despite not having the skills to carry them out, over-estimating her abilities. This would make her untrustworthy, but not necessarily a bad person, rather one lacking self-knowledge. Either way, she should not be relied upon.

Given that trustworthiness is relative to an audience, someone can easily be both trustworthy and untrustworthy at the same time – to different audiences. One obvious example is the spy. Suppose Alice works for the government of Freedonia, but also spies for Sylvania. She is committed to protecting the secrets of Freedonia, but is in fact leaking them to Sylvania. She is therefore certainly an untrustworthy Freedonian civil servant. However, she might still be a trustworthy spy for her Sylvanian bosses, to whom she also has commitments, which she is willing, able and motivated to carry through, including the commitment to break her commitments to Freedonia. She must pose as a trustworthy Freedonian without being one, and conceal her trustworthiness as a spy. Her *apparent* audience is Freedonian, but her *real* one is her Sylvanian spymasters. She may not have any pre-existing loyalties to Sylvania – the commitment may be reciprocal and transactional, rewarded by money, rank or medals (or even secured by blackmail). If she was a double-agent, the situation would become even more complex. The life of someone in such a position would be suffused in betrayal – a moral ambiguity that has driven an entire genre of fiction, exemplified by John le Carré's *A Perfect Spy*.

A "lovable rogue" who makes no particular pretence of trustworthiness may be perceived as more attractive than a hypocrite who pretends to be trustworthy and yet is not. In politics, this may present as a gothic style of populism in which relatively shady characters do not hide that they are on the make, but succeed by tarring their opponents as dishonestly occupying the moral high ground. "I'm on your side," they say to their audience with a wink, "and at least I am trustworthy about being untrustworthy." A legendary pioneer of such tactics was Senator Huey Long of Louisiana (Sanson 2006), and more recent Peasant leaders have used similar methods ("Trump is an asshole, but at least he's *genuinely* an asshole" – Shriver 2024, her emphasis).

SIGNALLING TRUSTWORTHINESS

It is in the would-be trustee's interest to be trusted (whether she is trustworthy or untrustworthy), so Alice needs to get the message out about her trustworthiness (whether or not it is true). Would-be trustees send *signals* of their trustworthiness to potential trustors exploiting a range of semiotic codes. Conscious signals include smiling, a warm manner, a firm handshake, smart suits, qualifications and credentials. Mastering a signalling system requires an understanding of what others are going to find convincing. Trying to signal trustworthiness on one's own terms ("take me or leave me") is probably a losing strategy. Luhmann wrote of "the personality which [the trustee] has presented and made socially visible" (2017, 43; also Goffman 1959); this needs to be tailored for the audience. Unconscious signals also play a part, such as not sweating, maintaining eye contact and using open body language. Adverts and public relations are elaborate, mediated signals, while a would-be trustee may submit to customer rating systems, or the audits and codes of conduct of institutions.

A signal of trustworthiness that is hard to fake is valuable for trustors and trustworthy trustees (Pentland 2008). In cases of subconscious signalling, signals can be hard-wired via human neurophysiology. In professions where strangers can be dangerous, such as prostitution and taxi-driving, experienced practitioners look out for tell-tale clusters of signs of untrustworthiness in clients (Gambetta & Hamill 2005). In game theoretic accounts of trust, signalling can ameliorate the global suboptimality of equilibria in cooperative games such as the prisoner's dilemma, where each player's utility-maximising strategy leads to a worse solution for everybody (Skyrms 1996, 56–57).

The signals sent out must be *legible* to the receiver (the intended audience). Bob must be able to parse Alice's signals, realise their import and understand them as genuine. This requires a shared view of the commitments and context, suggesting that a common core of culture or language will be an advantage in successfully communicating trustworthiness and acting effectively upon it.

Signals may mean different things to different audiences. Military insignia, for example, or countercultural regalia such as hippy or goth gear, may be reassuring to some people, off-putting to others. Tattoos signal a certain alternative *cachet*, including membership of particular groups (Buss & Hodges 2017), but can carry stigma with more conventional people (Broussard & Harton 2018). Those worn by Central American gang members, or the full-body *yakuza* tattoos in Japan are very effective signals of trustworthiness for the criminal audience, because they constitute a permanent and costly barrier against acceptance in wider society (Brenneman 2014, 114–116; Hill 2014, 238), signalling *untrustworthiness* to ordinary law-abiding people.

Signals need not be accurate. If Alice can send the right signals independently of her trustworthiness, then she can signal trustworthiness without actually being trustworthy. Being trustworthy becomes an extra cost. Therefore a successful signalling system needs to be such that it reliably flags lack of trustworthiness. The problem is that the moment a signal system becomes an object of scrutiny and a conscious factor in trustors' decision-making, then it can be faked, which allows trustworthiness to be signalled deceitfully. Crowdsourced online marketplace ratings, for instance, are effective in signalling (and incentivising) the trustworthiness of sellers, until the untrustworthy learn how to use bots to fake their ratings. This requires a response from the marketplace to prevent or disincentivise it, leading to an arms race between the platform and the untrustworthy (Malbon 2013). Social media accounts that disseminate misinformation often begin by spreading reliable clickbait about low-stakes topics such as football or *Game of Thrones*. Malicious software was recently inserted into an open source component of the Linux operating system by an apparently trustworthy agent who had provided hundreds of pieces of helpful code over the previous two or three years (Greenberg & Burgess 2024).

The requirement for signalling marks trustworthiness out from other kinds of virtue or ethical principle. Jesus preached that good actions should be anonymous: "when thou doest thine alms, do not sound a trumpet before thee, as the hypocrites do in the synagogues and in the streets, that they may have glory of men … let not thy left hand know what thy right hand doeth: that thine alms may be in secret: and thy Father which seeth in secret himself shall reward thee openly" (Matthew 6:2–4). When virtue is publicised, on this view, it loses its ethical character and becomes a type of self-aggrandisement.

On this measure, if trustworthiness is a kind of virtue, it is an unusual one (another example of a virtue with an ineradicable public aspect is honour). It has a clear goal that will directly benefit the virtuous person – that of being trusted. And to be trusted, she needs to signal, to blow her own trumpet (if not actually sounding one). Trustworthiness of necessity needs publicity. There is no point in someone making sincere commitments to an audience which lacks the awareness to take her up on them.

GENERAL TRUSTWORTHINESS

I have described trustworthiness as relative to a commitment, audience, and context. However, we sometimes make broader statements, such as "Alice is trustworthy" or "Alice is untrustworthy," derived from underlying regularities in her behaviour. If she generally does what she says, is willing, able and motivated to carry out her commitments for her audiences conscientiously, and doesn't dodge taking on commitments, then we can generalise from that virtuous behaviour to say she is a trustworthy person *simpliciter*. It will also mean that she avoids making commitments where she knows there may be problems, but we would not normally say of someone who is sparing in making commitments that she is generally trustworthy, even if she is conscientious in carrying them out on the rare occasions when she does.

General trustworthiness is a pattern of virtuous behaviour with respect to many other people and commitments. Whereas the individual manifestation of trustworthiness with respect to a specific commitment may be transactional, general trustworthiness is a positive character attribute that can explain (but not cause) specific instances of trustworthy behaviour. In this sense, it fits into the pattern of an ascription of virtuousness: "It is not merely … a claim about how things are with the agent and her reasons at the moment. It is also a substantial claim about the future (with respect to reliability) and, most importantly, a claim about what sort of person the agent is" (Hursthouse 1999, 123). However, such an attribute is always tested by new circumstances and challenges. Lord Jim, in Conrad's novel, is a fine, brave fellow whose character is unquestioned, until he abandons his ship during a storm, an action he can scarcely understand of himself (O'Hara 2007, 62–64).

Similarly, Alice is untrustworthy *simpliciter* if she is regularly cavalier about her commitments. Her general untrustworthiness is a negative characteristic, fitting into a pattern of ascriptions of vice.

CODA: TRUSTWORTHINESS, ASSOCIATION AND INDIVIDUALISM

Trustworthiness is a vital means of coalescing and binding groups, little platoons, associations and communities, depending on the size and coherence of the audience. Given an audience and a robust signalling system, trustees can furnish the conditions for collaboration, interaction and reciprocity, ensuring that the group is greater than the sum of its parts.

In an individualistic age, trustworthiness presents a problem, since the rationally self-interested individual would seem to gain so much more, at least sometimes, from betrayal of her fellows. However, the poverty of such an anthropology is revealed when we consider the extent of trustworthiness that is evident in almost all areas of life in Western democracies. This may not apply to some other types of society, or to socialist paradises. But most people are generally trustworthy, want to be so, and will try to help, to be reliable, to take on commitments and carry them out. Even hard-nosed businesspeople tend to be trustworthy, as McCloskey argued in full-on sarcastic mode:

> For instance, what do you suppose is the rate of "nonperformance," as the bankers call it, on big loans to business in the modern world? Ten percent? Twenty? I mean, these are nasty, dishonest people, these businesspeople. Well. It's in fact 2 percent. … Trustworthiness is a business virtue … . (McCloskey 2006, 350)

Trustworthiness echoes an imperative that older civilisations thought was evidently good, useful and important. Many moral ideas across various societies and cultures, from Ancient Egyptian *Ma'at* to Confucian *li*, from Ancient Greek *Dike* to Burkean order, from the *sensus communis* (community sense) of Immanuel Kant to Hegelian *Sittlichkeit*, are concerned with preserving the moral order and encoding intuitive fairness into social conventions and norms. Following such norms means respecting one's society and the members of one's community. We have no modern-day equivalent of these public and communal ideals, but trustworthiness approximates their results if enough people are trustworthy to enough of their fellows enough of the time.

4. Trust

In this chapter, I shall introduce what is involved when Bob trusts, or doesn't trust, Alice. If he trusts her, he believes that she is trustworthy, and thus that she is willing, able and motivated to carry out her commitments to him. Even given this very basic description, we can see that trust has a subjective aspect, and uncertainty stemming from Alice's free will. Furthermore, his trust in her can be high or low, in which cases his belief in her trustworthiness is strongly or weakly held. Add to this its social aspect rooted in forms of life, its contextual aspect, its hermeneutic aspect (interpreting commitments and the extent of the audience) and its communicative aspect (signalling), as well as the ambiguities sketched in Chapter 2. It is not simple!

Trust is also often reciprocal, and disseminated through networks of mutual trust and reliance. The picture is so complex, and the literature so vast, that a chapter of a book can only scratch the surface. I will therefore focus on the features that impact political trust and ideology in particular, making the simplifying assumption that, for heuristic purposes, an account of Bob trusting Alice will be sufficiently informative (and even that will be complicated enough).

The problem of trust is to ensure that trust is placed in all and only trustworthy people, so Bob's assessment of Alice must be accurate enough to weed her out when she is untrustworthy, yet not so rigorous as to miss an opportunity to collaborate by being over-cautious. Ideally, Alice and Bob will have an empathetic relationship, to enable them to judge each other accurately in terms of their willingness to commit and cooperate, but this may not always be possible. When the level of general *trustworthiness* within a society is high, trust is a broadly winning strategy.

THE PARAMETERS OF TRUST

Trust is an evaluative attitude that Bob has toward Alice's intentions and volition, and so is inherently uncertain. Institutions and practices have emerged to reduce this uncertainty. Trust is rarely entirely rational, and depends not only on Bob's assessment of Alice, but also his interpretation of what she is committed to.

As I did with trustworthiness, I will begin by stepping through the parameters of trust. I will focus on graded-trust as opposed to threshold-trust, as the former is conceptually prior.

Agents

Bob, the would-be trustor, could be anyone or anything able to form such an attitude toward another. Many have wondered whether artificial systems, agents or robots can trust (Taddeo 2010), but I will gloss over this question and take the paradigm case of Bob as a person.

The object of trust is also paradigmatically a person, but may be a system, role, institution, technology, even a piece of information. Bob's trust may be in Alice's trustworthiness, or in an artificial object's reliability, but I focus on trust-in-trustworthiness here.

Bob's trust may be in Alice the person, or *qua* a role she occupies, or both. If he trusts her as a person, he believes that she is willing, able and motivated to carry out the commitments she has taken on. In role-based trust, he has to believe that (a) the role involves carrying out commitments for him as audience, (b) the role gives Alice the resources to enable her to do so, and (c) she is willing, able and motivated to occupy the role. He will often take all these on trust from a third party with standing (for example, Alice's boss).

Bob also may occupy a role which influences whom he trusts. For instance, if he is interviewing candidates for a job, his role in the firm will limit the number of factors he can bring to bear on the decision. He may have personal reasons for trusting Alice more than Adele (for example, he may believe that Adele is likelier to demand flexibility in her working hours because of childcare responsibilities), but his role will not allow him to bring these to bear on the decision, which must be based on a strict set of fairness criteria.

Finally, it is also possible that Alice and Bob may be the same person – i.e. one can trust *oneself* to a certain degree. Bob the alcoholic may take the dog for a walk on a circuitous route that avoids going past the pub, because he doesn't trust himself to carry on past. He may signal a new level of confidence in his ability to resist temptation in future by taking the more direct route.

Commitments

Alice has made commitments that may be open-ended or specific, implicit or explicit, abstract or concrete. The question is what Bob makes of them.

The first point to note is that Bob has to *understand* Alice's commitments, and *interpret* them. They must have been communicated in some way, so that he has some expectations. He need not care about the full range of her commitments; what counts is *his interpretation of what Alice's commitments would mean in the specific context he is interested in.* For instance, he may not care about Alice the mechanic's full range of skills, but specifically about her ability to diagnose the warning light on his dashboard, and to cure the glitch. He may not care about Alice the MP's ideology in the round, but specifically whether she is willing and able to protect his job at a lossmaking factory. Both Alice and Bob will have their own interpretations of how the commitments apply in Bob's context, and if his trust is to be accurately placed, these interpretations must coincide.

Where commitments are attached to a role, Bob must interpret how the role applies in his context. Where Alice's commitments are specified by Carlotta, a third party, the situation becomes correspondingly more complex. Carlotta must have the standing to commit Alice to a role, and Bob must also believe, correctly, that she has.

The Degree

Graded-trust, as opposed to threshold-trust and placing trust, is not a binary, but a judgment about someone that comes in degrees. This may be expressed in different ways. We might assign graded-trust a real number between 0 and 1 to give us a continuous scale, or we might have a small qualitative set of discrete grades: {complete trust, substantial trust, reasonable trust, neutral, little trust, negligible trust, zero trust}.

No doubt there are psychological experiments that could help with this, but we don't need to solve the question here (Eriksson & Hájek 2007; Huber & Schmidt-Petri 2009; Leitgeb 2017). We simply need to note that we are able to make comparative judgments that can't easily be captured in binary form. Bob would trust Alice with £100 but not £1,000. Bob trusts Alice more than Adele. Bob lends to Alice at 5% and Adele at 10%. All this shows how graded-trust is at least as important as threshold-trust, even if the latter is more prominent in commentary.

We often write that "Bob trusts Alice" to mean "the degree of Bob's trust in Alice is high," and "Bob doesn't trust Alice" to mean "the degree of Bob's trust in Alice is low;" I have used and will continue to use that shorthand on occasion in this text.

The Audience

Alice's commitments are made to an audience. It follows that Bob, if he is to rely on her, must believe that he is a member of that audience (or, where he is acting as an agent or broker for others, that his clients are members of her audience). If his belief is incorrect, he may be placing his trust in someone who has made no commitment to him at all.

Bob may have a view of Alice's trustworthiness, but realise that he is not part of her intended audience. She aims her skills and goodwill at other people than him. Perhaps he is a crook and she is an upright policewoman. Perhaps he is wealthy, while she is a lawyer who specialises in advising those in poverty. As a man, perhaps he knows he cannot benefit from her medical expertise on women's health. As a highly educated and qualified person, perhaps he has no need of her teaching skills. Perhaps he is a couch potato, while she is an elite sports coach. Perhaps he and Alice are competitors providing similar services; he recognises her trustworthiness, but will never need it.

In all these cases, he can, of course, still exhibit trust in her, for example by recommending her to others, but direct behavioural evidence will be thin, because he will rarely have an opportunity to *place trust* in her, despite his positive views.

The Context of the Trusting Encounter

Alice's commitments are relative to a context; they only apply in certain circumstances (e.g. she may only be committed to paying customers, or during office hours, or she may need specialised resources available). Bob's request for help from Alice takes place in a context on which each has a perspective. It is inhabited by Bob, and in it he will have some sort of issue that he believes Alice could help with, were she to commit to him. When he requests help, Alice becomes aware of it, and must assess it as falling within the set of contexts within which her commitments apply.

The context is often social, one of institutions, traditions and practices. Sometimes, it is so pervasive that neither Alice nor Bob will see it as an occasion for *trust* at all. The practice is carried out as a form of life in which both are well-schooled and habituated – this is just what we *do*. Yet the practice, if it is unpicked, might include hidden risk for Bob, and an outsider might therefore infer an implicit attitude of trust of Alice.

Warrant

Bob's belief about Alice doesn't come out of the sky. It rests on some kind of ground, which we might call a *warrant*, or explanation. This need not be convincing in an individual case, but across wider society, enough warrants need to be broadly accurate for the practice of trusting to be functional. Bob assesses Alice's trustworthiness as high or low, and the warrant should explain why that is so.

Not all beliefs are rational or sensible. The warrant for Bob's assessment of Alice's trustworthiness may be silly, or even biochemical. It may be a good or bad *reason*, or it may simply *cause* his belief. In general, his trust (or otherwise) in Alice may depend on (be explained by, be caused by) a wide range of phenomena (Lenard 2012, 15). It can be manipulated, and deeply irrational ("I trusted her because of her big blue eyes"). It can also be highly rational, and human society has developed a number of institutions, formal and informal, to facilitate accurate judgments (global trust). Alice's signals of trustworthiness are important, but may not be the only part of the story. Some signals, given and read subconsciously, cannot be cited by either trustor or trustee.

Examples of factors that the warrant may involve include experience and memory of previous interactions, judgments of character, qualifications and credentials, social reputation, incentives and sanctions, joint membership of a clique, peer pressure, power relations between trustor and trustee, social norms, the ambient level of trust in society, and individual predispositions of the trustor to trust. Some of these depend on the personal acquaintance of local trust, others on the socialised mechanisms of global trust. Furthermore, many will depend on Bob's trust for others; for instance Bob's trusting Alice on Adele's recommendation depends on how far he trusts Adele as an advocate.

Bob has an interest in being able to demonstrate good reasons for trusting Alice, because he will sometimes need to justify his trust not only to himself, but socially. If he has been a fool to trust, he may lose social standing, and be unable to claim compensation or even sympathy from his fellows (Luhmann 2017, 29). Any *general* descriptive account needs to be neutral between sensible and idiotic warrants. Even some of the more rational methods available to a trustor can be subverted if the trustee is devious enough: qualifications can be faked, recommender systems spammed, reputations manipulated. It is easier to trust a winner, so Bob is likelier to trust financial advice if Alice has an expensive suit and a flash car – a fact well-understood by con-artists for centuries.

If trust had always to be based on credible evidence, it would never get off the ground. A normative account of trust demands rational and evidence-based warrants, in the usual case. Many commentators reserve the phrase "warranted trust" for precisely such circumstances. The risk of privileging rationally-warranted trust over other kinds is that there is not enough of it around to oil the wheels of social life (Hollis 1998, 9–10), or to enable interactions with strangers in complex and dynamic societies (Uslaner 2002). Trust cannot be a wholly rational, evidenced matter, and so must rest at least partly on a learned but potentially arbitrary pattern of habits, norms and virtues contingently expressed in society, what Burke called prejudice, "the general bank and capital of nations and of ages" (1999a, 182).

Hence my preference for a descriptive account over a normative one. If we have hard and fast rules about when trust is rationally warranted (or morally good, for that matter), we suffer the consequence that we struggle to explain how trust even starts. Better to take as our starting point the evident fact that trust is widespread and broadly beneficial, and try to refine our account to encompass problems that might arise. Some are pessimistic that there is no way to rely on the practice to keep us secure (Williams 1988), but it seems the only feasible route forward.

Many theories suggest that we trust because we are better off when we do. This may indeed describe a winning strategy in certain types of society, but it's rarely the *reason* that we trust. While I am unconvinced by psychological experiments in this field, there is much evidence that, under laboratory conditions, people come into social situations with an intention to trust and be trustworthy, even when the setup supports anonymity or screens off useful reputational information. Even the more realistic strategy of "constrained maximisation," that it pays to trust/be trustworthy only with those who agree that it pays to trust/be trustworthy (Gauthier 1986), still seems unrealistically calculating (Hollis 1998, 95–97).

Luhmann made an intriguing and convincing objection that "treating the process of trust as an analogy of cognition – trust is only right when it is objectively justified – corresponds to a very widespread prejudice which is that stable structures within people could not be founded on unstable structures in their environment" (2017, 87). There is a lot to unpick in this, but his basic point is that trust helps stabilise expectations to enable richer interactions. The very fact that it is known that trust is insecure enables us to adapt and adjust when it goes wrong. We *expect* those we trust to be trustworthy, but are *simultaneously* prepared to be let down, as this is the nature of trust. We could not do this if trust required rational justification each time.

A Note on Trustless Trust

In many trusting interactions, Bob and Alice engage to determine the appropriate contexts for Alice's intervention, and the sorts of outcomes that will help Bob. The context can be manipulated by both parties, by Alice to make it more favourable to her expertise, and by Bob to make Alice's intervention more likely to benefit him. In the case of trustless trust, Bob's aim is to so reduce Alice's degrees of freedom that she has minimal opportunity to use her discretion. In that case, he has reduced the uncertainty that bedevils most trust judgments. Trust has not been banished, however. Trustless trust means that his trust for her is high, his *warrant* being his trust in the constraints that have been put in place.

LACK OF TRUST AND MISTRUST/DISTRUST

Trust can fail, can be absent, or can present as *distrust* or *mistrust*. Lenard put trust and distrust on a continuum (2012, 14), while Luhmann wrote that in a trustless world we "would not even be capable of formulating definite distrust and making that a basis for precautionary measures, since that would presuppose that [we trust] in other ways" (2017, 5). Distrust/mistrust is intimately, internally connected to trust, but the way we understand it depends on whether we are thinking about threshold-trust, or graded-trust.

Not Trusting

Let's begin with the negation of trust: Bob doesn't trust Alice. On the graded-trust model, this is best understood as meaning that Bob has no view of Alice's trustworthiness. For example, if he needs a plumber and is unaware of Alice's plumbing prowess or lack of it, then he cannot form an opinion. Perhaps he hasn't even heard of her.

A lack of trust may also involve a negative judgment. In the threshold sense, when Bob doesn't trust Alice, trust is below whatever threshold he deems relevant. Sometimes the threshold is actually referred to: "I don't trust Alice enough to lend her £1,000." Bob may refer to it idiomatically or jokingly, as when he says "I don't trust Alice as far as I could throw her." He can also move from one sense to the other, with a risk of paradox, as when he says "I don't trust Alice at all, and I trust Adele even less," switching from the threshold sense to the graded sense in the course of his sentence.

Mistrust and the Lack of Trust

Beyond the lack of trust, we talk of mistrust and distrust. These may be synonymous, but some make a distinction. For Lenard, distrust is a firm belief in the untrustworthiness of a would-be trustee, whereas mistrust is cautious agnosticism, open to the possibility of positive evidence (2012, 55–61). The agnosticism that Lenard referred to as "mistrust" plays an important role in her theory of democracy, but I am not sure it is helpful to characterise it in this way. It seems more like a lack of trust owing to lack of evidence, which doesn't deserve a negative label.

I will not make a firm distinction between the two, although there is often a sense that distrust is indeed somewhat stronger than mistrust, where the distruster either has a more fixed or a more general attitude. But "mistrust" rarely expresses agnosticism, and I'm not sure there is a substantive difference between the two. As it may be marginally more generic that "distrust," in general I will use "mistrust" in this book.

If Bob mistrusts/distrusts Alice, then he *believes her to be untrustworthy*. His mistrust, like his trust, rests upon his subjective interpretation of Alice's attitude to her commitments, and so may be misplaced just as his trust may be. If he mistrusts Alice, he actively believes that she has taken on commitments that she is (i) not willing, or (ii) not able, or (iii) not motivated to carry out.

Mistrust is therefore distinct from a *lack* of trust. This carries over to trust-in-reliability (*contra* Hawley 2019, 4); one can trust an object (believe it reliable), mistrust it (believe it unreliable), or not trust it (have no relevant belief).

Mistrust has warrants of its own, some of which are directly to do with the would-be trustee, while others may be present as background beliefs. For instance, if Alice is a doctor, Bob may mistrust her because he has new agey views of sickness and does not believe that modern medicine is efficacious. Or, if he has confidence in medical practice, he may still not believe that Alice has the ability to occupy the role of doctor successfully. Thirdly, he may be prejudiced – for instance, he may not believe that women should be in practice. Or he may think that Alice is an effective practitioner, but may mistrust her for other reasons – for instance, he may believe her to be a blackmailer who will take advantage of confidential information. Or he may trust Alice in all these ways, but also be aware of a conflict of interest. For instance, if she were both a doctor and a prominent member of an authoritarian governing party, he may be concerned that she might use his role as political dissident against him. Various factors may be in play, and the question of whether Bob mistrusts science, medical research, the practice of medicine, the technology of medicine, the pharmaceutical industry, doctors in general, or Alice in particular may not even have a determinate answer. He may simply be in denial about his health.

It has been argued that mistrust is a more difficult and expensive attitude to carry off than trust (Luhmann 2017, 79–84), which is probably a good thing. Mistrust can be a cynical reflex, as if trust were for the gullible, credulous and foolish. "Expressions of mistrust that are divorced from action come cheap: we can assert and rescind, flaunt or change, defend or drop attitudes and expressions of mistrust without changing the way we live" (O'Neill 2002, 14). Furthermore, mistrust, unlike trust, is not self-correcting. Willingness to place trust will tend to result in more interactions, and therefore more learning about trustees' trustworthiness. We can learn even from trust that has been badly placed. However, where mistrust prevails, there will be fewer opportunities to correct initial misjudgment (Lenard 2012, 57–58).

General and genuine mistrust, where even basic trust in the environment is lacking, is disabling. Without a sufficient foundation of belief in the trustworthiness of others and the reliability of the environment, a person will find it almost impossible to plan and act. "The outcome is persistent existential anxiety … a state of mind which could best be summed up as existential *angst* or *dread*" (Giddens 1990, 100, his emphases).

It is hard to react even to what appears to be a genuine crisis or failure of trust with a blanket adoption of mistrust. While we are likely to revise our current practices, and perhaps increase mistrust, we generally try to reconfigure our new surroundings with more realistic attitudes. The new normal might be more transactional and risk averse, as we seek out interactions that do not require elaborate trust mechanisms or rest on now-disrupted patterns of behaviour (Govier & Verwoerd 2002; Reemtsma 2012, 33–39; Goodwin et al 2023). The victims even of large disruptions, such as natural disasters, wars and enforced migrations, can preserve agency and social support through such reconfiguration, building on whatever positive experiences may have followed necessary episodes of collaboration (Hall & Werner 2022), perhaps with tighter in-groups and suspicion of out-groups (Cassar et al 2013; Kanal & Rottman 2021; Ali et al 2022; Renkens et al 2022; Muminovic & Edfendic 2023). That does not necessarily mean that in-group trust increases (Lewis & Topal 2021), but it may slow its decline.

The Interplay of Trust and Mistrust

Trust and mistrust often work in tandem to manage conflicts of interest. The game of poker is built around it; players trust each other not to cheat, but also expect them to lie about their cards. A negotiation involves keeping information back from counterparties, because it depends on information asymmetries, lying to each other while working together to find a win-win. Policymakers allow access to trusted business lobbyists, because businesses tend to contain far greater expertise on whatever is being regulated than the policymakers themselves. They trust the lobbyists to inform them of the practicalities of regulation, but are also fully aware that the lobbyists are also keen to promote their own company.

Where lies the boundary between cheating and kidology? It is impossible to say, and that is part of the game too. No wonder that personal connections between policymakers and lobbyists are valued, so that judgments are based on rich evidence. Even if their roles place them in opposition, Bob the policymaker may trust lobbyist Alice at a personal level, helping him feel more secure. On the other hand, this can lead to unfair outcomes; Alice and Adele may both lobby Bob, but if he is personally acquainted with Alice out of role, and doesn't know Adele very well, he may favour Alice's opinion, which would seem unjust to Adele (and may not be in Bob's interests either).

ALIGNING TRUST AND TRUSTWORTHINESS

Granted that trust should not be doled out as if it were going out of fashion (even though it is), what we really want to do is align it with trustworthiness. Those who believe trust is a vulnerability tend to exaggerate the challenges of this, worried by the covert, occult nature of the trustee's real intentions.

However, the world doesn't work like this, and mechanisms have evolved to help us achieve alignment. The would-be trustee's signalling has been discussed, but this won't help dissipate the trust sceptic's concerns, as signals may be falsified. But would-be trustors and trustees can also collaborate to reach a *modus operandi.* In particular, the trustor can use constraints and incentives to change his assessment of the likelihood of the other carrying out her commitments, while dialogue and negotiation can bring the two to a clearer understanding. Social norms prescribe regularities in behaviour that are helpful too.

Sanction and Incentive

Uncertainty, vulnerability and risk can be reduced if the trustor can engineer the context in his favour. Bob can deploy both incentives and sanctions, so that, while Alice is still free to meet her commitments or not (uncertainty is not banished), he can be a little more confident that she will work to his benefit. Applying incentives and sanctions changes the context of Alice's commitments, and may therefore give Bob superior grounds for confidence in her willingness, ability and motivation to carry them out. In the limiting case, the result is trustless trust, where Alice's room for manoeuvre is minimised as far as possible, but this is clearly wasteful of the skills and expertise she could use were she granted discretion.

It is important that Alice perceives the change in context. Suppose Bob uses a surveillance camera to monitor her performance. If she is unaware of it, she will perform as well or as badly as if she wasn't monitored. At best it may give Bob some evidence to criticise Alice *post hoc*. If, however, he warns Alice of its presence, then she is aware of the change in context, and knows that poor performance is likely to be spotted. Now, Bob's trust in her to carry out her commitments in the changed context is likely to be higher than when unmonitored.

Note that the *context* has changed, *not* Bob's graded-trust view of Alice. He may reason that "if the camera wasn't there, she'd go back to her old ways," or alternatively that "now she has seen I've got my suspicions about her, she'll buck her ideas up even if I don't watch her like a hawk"

Focusing on carrot rather than stick, Bob can pay commission, or dangle inducements such as the possibility of further work in the future. He can monitor her performance more or less closely, giving him more information about how she is meeting her commitments. When trust is hedged around with this sort of mechanism, Bob both gains and loses. He gains greater confidence that Alice will carry out her commitments, but monitoring involves expenditure of resources which offsets some of the gains. Global trust institutions have evolved to incentivise, sanction and monitor, benefiting from economies of scale, providing certification services to Alice at a lower cost to Bob.

Dialogue

An even more important mechanism for supporting trust is dialogue. Bob is not (always) a passive receiver of Alice's commitments, and in many, perhaps most, cases they can create a mutual understanding of her commitments to him.

Alice's commitments need not be take-'em-or-leave-'em. In many cases (especially where her commitment is not pre-existing), Bob can engage in dialogue with her, ask for her help, tell her exactly what he needs, and negotiate about her services, while she in return can explain her capabilities, and alternative pressures on her time and resources. Where the arrangement is reciprocal, the negotiation will go both ways. Dialogue can reassure Bob that Alice's actions within his context will be what he requires. Face-to-face dialogue is a rich seam of communication often overlooked by theorists who like to emphasise the trustor's vulnerability. In order to be trusted, Alice can open herself up to Bob, get to understand what he needs and expects, winning trust by being warm and open (Giddens 1990, 121). It also, as a social interaction, enables Bob to move beyond heuristic warrants to something more knowledge-based and discriminating.

Trustworthiness is an other-directed project and, all things being equal, selfish people will struggle to present themselves successfully as trustworthy. Dialogue does not prevent misunderstandings or betrayals, but it is a good means of testing the commitment of the other. A rich conversation may be all Bob needs to expose Alice's untrustworthiness.

Norms

Social norms are also valuable. These conventions and regularities enable people to sustain expectations about the likely future behaviour of others. Trusting arrangements can leverage these; in effect, the trustor trusts the trustee to stick to the norms that obtain in their society. A trustee's failure to follow some norms (for example, wearing non-standard business apparel) could signal to a trustor that she is a proud nonconformist, who disdains norms. If so, that could cause trouble for the trustee, because how can the trustor be sure that she will live up to his expectations? As we will see in Part III, many radical thinkers disapprove of conformism, erroneously thinking that it indicates a lack of imagination. Nothing is further from the truth – the person who lacks imagination is the nonconformist, unable to empathise with those who are nonplussed by her behaviour.

Sociality as a Resource

Constraints, sanctions, incentives, dialogue and norms are not only direct influences on behaviour, but also help foster the social understanding of what trustworthiness consists in in the relevant context. The social repertoire may include contracts, regulations, punishments, ideal consensus, promises, custom and tradition, natural law, the agreements of rational people, up to and including faith in an all-knowing God. These are important, not only changing the cost/benefit analyses of would-be trustors, but also signalling social approval or disapproval of certain actions and relationships (Luhmann 2017, 41).

None of this social repertoire is sufficient to *make* someone carry out her commitments if she decides not to. Free will prevails. Neither is it necessary. Just because Bob could take Alice to court if she botched the repair of his car, it doesn't follow that she would have produced shoddy work without the legal framework.

Nevertheless, if we ignore the rich social context that supports accurate trust judgment, then placing trust, especially in reciprocal arrangements, looks harder than it actually is. In a common enough setup, Alice and Bob agree to cooperate, where Bob will first help Alice, and then Alice help Bob. As both Hobbes and David Hume pointed out, while Bob knows he needs to help Alice in order to receive her help, once he has delivered his side of the bargain, Alice has no incentive to continue cooperating. If she has gained all she can from Bob, then why should she spend her own time and resources helping him, when she would receive no further benefit? Worse, if Bob suspects Alice of planning to betray him (since he can reason about the future situation just as well as she can), then he has no incentive to help her in the first place, so the cooperation never gets off the ground at all. If we take this narrow view of self-interest, we are in trouble (Hollis 1998).

This is more thought experiment than daily experience, but it helps us see the spurious attraction of trustless trust. One cannot decide now how to behave in the future; one can always change one's mind (Haugeland 1998, 341), and it may appear to be necessary to use technical means to deprive oneself of choice altogether to both demonstrate and adhere to a commitment. With distributed ledgers and smart contracts, what is agreed now is made to happen in the future by the immutability of the blockchain.

The fallacy in the reasoning is in the abstraction of the example. In the Alice/Bob cooperation problem, it is simply not the case that Alice would receive no further benefit from helping Bob. The continuation of their partnership, the bolstering of her own reputation, the cosseting of her conscience, the avoidance of an unpleasant confrontation, the satisfaction of a job well done, and the pleasure that empathy provokes in enterprises from which we do not directly gain are all indirect benefits to Alice of continued cooperation.

In any case, these would only be "needed" if there was a chance that Alice would disrespect her own commitments sufficiently to renege upon them. It is often said that virtue is its own reward. While this is usually intended ironically, it is actually true, and one can see it in daily life. Many philosophical accounts suggest that individuals police their own actions; Adam Smith argued that one acts as if being observed by an impartial spectator, Kant that failing to fulfil one's commitments puts one in conflict with one's own will, Heidegger that one's Being demands and is partially constituted by one's commitments (Crowell 2022). A person who aspires to be trustworthy takes commitment seriously, and trustors can leverage that maxim.

Negotiation and clear communication are valuable for addressing hermeneutic problems, although they cannot be guaranteed to be successful. "It can be difficult for someone entering a new context to appreciate what the local conventions and practices are, and so to appreciate what commitments she may be taken to have incurred" (Hawley 2019, 125).

> Controversial empirical studies of levels of "generalized trust" in multicultural neighbourhoods have suggested that trust levels in such areas are low. [This may be down to low incomes, but] a complementary factor may be mutual misunderstandings, or differences of opinion, about what actions and behaviours trustworthiness requires. This may be as mundane as expectations about what to do with household rubbish, noise levels at different times of the day, or responsibilities to children out playing in the street. (Hawley 2019, 135–136)

This is one reason why groups and associations are important for trust; joint membership may help align Alice's and Bob's understanding of her commitments. Membership may provide some scaffolding assumptions for them to share, while the gatherings of the association will also provide opportunities for them to meet, directly or indirectly across the network of members. More formally, roles may be designed or redesigned by the association so that Alice's discretion and Bob's interests are optimally aligned.

RISK, BELIEF AND ACTION

The ambiguity between the attitude of graded trusting and the action of placing trust is not always noticed and rarely discussed as significant (Rotenberg 2010 is an exception), which may help explain why trust looks to some like an Achilles' heel. For instance, Patent and Searle (2019), reviewing the literature, discuss three "requirements" for trust: dependence of trustor on trustee, vulnerability stemming from the losses that can result from trusting inappropriately, and uncertainty about the trustee's intentions. This triad emphasises the trustor's problematic weakness. However, clearly we can trust without these requirements being very salient. If we describe trust *as* a vulnerability, then it *becomes* a mystery. But seen in the round as a *form of life*, the vulnerability might be of vanishing significance, as when Bob gets on the bus without knowing whether the passengers are assassins.

As trust is a belief, like other beliefs it need not be acted upon. "To entrust is intentionally and usually formally to hand over the care of something to someone, but trusting is rarely begun by making up one's mind to trust, and often it has no definite initiation of any sort, but grows up slowly and imperceptibly" (Baier 1994e, 105). This quote expresses well some of the relations between trust as an action and trust as an attitude, but unfortunately fails to make the crucial distinction between them.

When Bob places trust in Alice, then he behaves as if she will fulfil her commitments (Holton 1994), which may induce dependence, vulnerability and uncertainty. Placing trust means taking a risk (not necessarily a large one). It is the *action* that makes Bob vulnerable. But if he, while rating her trustworthiness highly, never actually places trust in her, then he *cannot* be dependent on her, and is not vulnerable to her betrayal. If his trust is strong and meaningful, as with a family member or close associate, then, even if his trust does technically induce vulnerability, it would be a mistake to exaggerate its importance (Hawley 2019, 7). And, in contexts that are not time-critical, he may pursue dialogue and tailoring of context to reduce any vulnerability still further.

It is perfectly consistent to trust someone and place no trust in them, take no risk on them, most obviously when we do not require their services (though one would be prepared to take a risk in the right circumstances). And because trust as an action is somewhat divorced from trust as an attitude, it is also possible to place trust even when the level of trust is low, as I shall discuss in the next subsection.

Placing Trust with Low Trust

Luhmann wrote that "if chaos and paralyzing fear are the only alternatives to trust, it follows that man by nature has to bestow trust" (2017, 5, and also 30; O'Neill 2002, 12–14). This is absolutely true. One forms graded-trust judgments about the contexts and the risks of placing and not placing trust, and in many cases, placing trust is the least worst option. One's level of graded-trust may be very low – little faith that the trust will be rewarded. However, the uncertainty that characterises trust remains; one can never be sure that a trustee will deliver, but by the same token, neither can one be sure that an untrustworthy trustee won't.

Imagine Bob in a desert country. He does not trust the water company very much, but, if there is no bottled water, he will have to drink from the tap. He has to place trust (or if you prefer, to rely), even in the absence of positive belief. He might not want to lend Alice £10, but it could be embarrassing to show his lack of trust in her (in front of friends, say). He may be quite reluctant to place trust, but have no choice as a matter of politeness.

In another type of case, the trustor tries to engineer an outcome which will be facilitated by trust, showing faith in the ability of the would-be trustee without necessarily having a strong positive belief behind it, to serve a higher purpose. This helps the trustee learn trustworthiness. Examples include setting stretch goals for learners. Bob the parent, teacher, or clergyman may want to encourage Alice the daughter, pupil or delinquent along the straight and narrow path, showing trust in her performance without necessarily believing she will succeed, or even intends to succeed. Nevertheless, he behaves as if she will, because the very act of trust makes her trustworthiness more likely, not only now, but also in the future. Bob deliberately lowers the threshold above which he will place trust, because he is taking into account a future relationship beyond the possible betrayal in the present. As we will see later, this has been called *altruistic trust* (Mansbridge 1999).

This may seem abstract and even paradoxical, but is hardly unusual for anyone in a position of responsibility, and not only in Western societies. In the 18th century Chinese novel *The Story of the Stone*, the dissolute Xue Pan asks to leave home to make his way in the world. His mother and sister debate the merits of allowing him to go:

> "It's good that he should want to occupy himself with something serious at last," said Bao-chai. "The trouble is, of course, that however fair-sounding his intentions may be now, once he's outside he may succumb to his old weaknesses again, and then it will be that much more difficult to control him. However, I suppose that is a risk one has to take. If he is really going to reform, then this experience may prove a life-long blessing. If he is not, I don't see there is very much left that you can do. After all, there is only so much one can do for another person: the rest must be left to Heaven. Pan is a grown man now, Mamma. If you maintain that he is too ignorant of the world to be allowed out into it, you are not going to make him any less ignorant by keeping him shut up indoors. And since he is being so reasonable for once, I should make up your mind that you are going to lose eight or nine hundred taels and hand it over to him to see how he manages. After all, he will have someone from our own business helping him and they may well feel some compunction about cheating him; so it's by no means a foregone conclusion that he will lose it all. And for another thing, away from home there won't be those worthless companions of his to egg him on; nor, on the other hand, will he have anyone to fall back on if he gets into trouble. He will have to fend for himself: eat when he can and go hungry when he can't. And when it dawns on him that he is on his own and has no one he can lean on, he may begin to behave a bit better. Surely it's worth trying?" (Cao 1977, 450, his emphases)

Where Bob occupies a role, the paradox of placing trust with low trust disappears readily. Bob in his own right may have beliefs that he is not allowed to act upon as the occupant. A teacher is simply banned from treating some of his pupils as destined to fail, even if he thinks they are; an employer cannot behave as if his employees are workshy; a clergyman is required to look upon everyone he comes into contact with as if they could plausibly go to Heaven; a football referee must bring on the stretcher for a writhing, screaming player he knows perfectly well is uninjured. Trusting in-role is unavoidable, independent of the actual beliefs of the role-holder. The role carries with it a template for action, so its occupant must place trust in certain ways that may on occasion diverge from his actual assessment of the trustworthiness of the trustee, or face discipline.

Hence, a low level of trust is compatible with placing trust. Placing trust primarily requires balancing risks, and where all options are risky, one may have to take a punt. We should not deduce a high level of trust from that. Given a meaningful choice or high risks, it is easier to place trust consistent with our trusting attitudes. When risks are low or choices limited, trust may have to be more promiscuous.

When Trust Fails

Placing trust brings the potential for tangible gain. If Bob places trust in Alice, he can go about the rest of his business on the assumption that she will fulfil her commitments. The corresponding risk, of course, is that she will fail to prove trustworthy. This is unavoidable, and ultimately, placing trust is an inevitable part of life. "The alternative is to live in a state of permanent uncertainty and to withdraw expectations without having anything with which to replace them" (Luhmann 1988, 97).

Where trust fails, it may be on a number of grounds. Even if Alice is trustworthy with respect to her commitments to Bob in his context, it is always possible that she has been waylaid by problems beyond her control. Congested traffic may have kept her from an appointment; illness may have prevented her from acting; inaccurate data may have stymied her calculations.

Alternatively, Bob's judgment of Alice's trustworthiness may have been inaccurate. The failure may be down to misjudgment of any of the many parameters of his trust or her trustworthiness. Bob and Alice may disagree about whether he is part of her audience, or whether her commitments apply in his context, or what they amount to in his context (as emphasised by Hawley 2019). Maybe Alice takes her commitments too far and illegitimately interferes in Bob's life (Baier 1994e, 101). Maybe Alice has not actually accepted the commitments at all, despite what Bob believes. Such missteps may be down to failure of communication, or it may be that Alice is not trustworthy in the relevant respects, or even that she is actively untrustworthy. Alternatively, Bob may be unreasonable in his expectations.

Where Alice occupies a role, her interpretation of the role and of what it entails in Bob's context must coincide with his for a successful conclusion. Where the role allows discretion for its occupant, the interpretative ground can be quite shaky. The role may give Alice too much power over Bob for his liking (Baier 1994h).

The Co-creation of Trust and Trustworthiness

There is a saying that trust arrives on foot and leaves on horseback. The picture is very simple: trust has to be earned through a long and painstaking series of trustworthy acts, and only then will one be trusted. However, it may all be lost in an instant – one slip-up and trust will be gone forever.

One of the many problems with the picture is that it suggests a determinate direction of causality: trustworthiness causes trust. Alice goes around being trustworthy, and eventually, through signalling and reputation systems, this reaches the notice of Bob, who then is prepared to place trust in her. This describes rational trusting in shallow game theoretic terms, but has obvious difficulties.

First, there is a bootstrapping problem: how does Alice begin to signal her trustworthiness convincingly? If people have placed trust in her already and she met her commitments, she can signal this and through further collaborations incrementally improve her reputation. But if she is new, without reputation, how does she begin to collaborate in the first place (on this model)? This is a serious objection, because the bootstrapping problem is easier for the untrustworthy to solve. By faking signals, untrustworthy Adele can build up a positive reputation. Trustworthy Alice has no such resources.

Trust can even be a *requirement* for trustworthiness to be displayed at all. The successful exercise of expertise depends upon the expert being given sufficient respect and deference (Agnew et al. 1997). The Sherlock Holmes comedy *Without A Clue* rests on the amusing premise that the real genius of the partnership is Watson. Holmes is an alcoholic actor employed to distract the police, press and public with his absurd clothes and gnomic utterances, allowing Watson to do the detection undisturbed while Holmes gets the credit. When, inevitably, they fall out, Holmes is laughably incompetent, but neither does Watson flourish, because no-one answers his arcane questions, or takes his deductions seriously.

Secondly, more importantly, trust and trustworthiness evolve together, not only over time, but within individual relationships. Writing a contract follows discussion and negotiation of what can be done at what cost and benefit. Through this process, each party learns something of the other, not only with respect to the matter in hand, but also more widely as they see the constraints on the other (for the importance of learning trust, Luhmann 2017, 31, 71).

Hence, trust begets trustworthiness. Trusting is an informative act. It teaches the trustor about the trustee's trustworthiness, and instructs the trustee what trustworthy action consists in, so they can learn through doing what trustors expect of them. By these means, trusting brings newcomers into moral or cultural communities (or brings transgressors back – Lenard 2012, 32). Trust is a means of showing that barriers of incomprehension between diverse people can be brought down. Going so far as to place trust, to take a risk, is a demonstration of faith.

In many professions and religious callings, trustors must be prepared for the disappointment of being let down. Those without supportive networks, not well-socialised, perhaps leading chaotic lives, must somehow be brought to understand what wider society expects of them, and enabled to meet these expectations. This cannot be done by reading books or listening to sermons. A would-be trustee must learn to recognise the disappointment and resentment of those she has let down, and ideally to feel it, empathise with it, and take it as a negative sign and a stimulus from which to learn trustworthy behaviour and the avoidance of temptation. On the trustor's side, for a betrayal to be met by disappointment, he must have had some positive expectations to disappoint. If Bob trusts Alice only grudgingly and warily, then she will struggle to learn to associate trustworthy behaviour with social success and acceptance.

Those who are perceived to place too much trust are often dismissed as "do-gooders." This is sometimes harsh, but trust cannot be unconditional. There must be a downside to untrustworthiness in the setup, there must be sanction available, because otherwise how would errant trustees learn through their own disappointment? Indeed, not only must trustees learn, but the rest of society must be protected from them while they are learning.

Conversely, mistrusting can cause untrustworthiness (Luhmann 2017, 82). If Alice comes to feel that her husband Bob never trusts her when she spends a night away, why, she may ask herself, should she not enjoy herself if she gets the opportunity, if she will get the third degree anyway? If young Alice feels that the police, her teachers and her neighbours all expect her to break the law and behave antisocially, what, she may ask herself, does she have to lose by living enjoyably down to her reputation? Those who don't *feel* trusted are more likely to do the minimum, watch the clock, apply the rules inflexibly. "Sticking to the letter of a promise rather than its spirit can often be a form of untrustworthiness" (Hawley 2019, 10).

In general, trust, trustworthiness and being trusted are delightful corollaries of belonging, and, in intimate circumstances, love. A straightforward rational or functional account will miss this; statements such as "personal trust is only formed where it is needed" (Luhmann 2017, 50) seem at odds with common experience. The causal links between trust and trustworthiness (and other social attitudes too) are myriad, and central to the development of abilities to associate, socialise and sustain communities.

GENERAL TRUST AND GENERALISED TRUST

So far, the discussion in this chapter has been about trust in specific circumstances. Of course, placing trust must always happen in a context, but we also make more general trust judgments, analogous to the character trait of general trustworthiness introduced in Chapter 3. Some have introduced the idea of generalised trust, a moral imperative to trust people in general (e.g. Uslaner 2002, 26–38), good-humouredly relying on the beneficence inherent in human nature (Luhmann 2017, 5). Despite the generous sound of this, it is an unhelpful normative ideal. I will distinguish *general trust*, a widespread and benign phenomenon, from *generalised trust*, whose problems will become evident when we go on to discuss politics in Parts II and III.

General Trust

We can use the term "general trust" in relation to a character trait of a trustor, or a more specific relationship with a trustee. The character trait, *generally trusting*, is a greater than usual tendency to trust and be trusting, a psychological tendency to rate others' trustworthiness on average more highly than others do, and/or having a lower threshold for placing trust. This may be a sensible way of going through life, at least in a bourgeois democracy, although some have argued that those whose trust exceeds social norms sometimes have psychosocial problems (Rotenberg et al. 2005), and they are certainly at greater risk of disappointment.

The converse, general mistrust, involves being generally mistrustful, perhaps a sensible strategy in a dangerous and risky *milieu*. However, those who are generally mistrusting, particularly children, also have difficulty getting along with others (Rotenberg et al. 2005). Departing from basic trust norms in either direction can lead to rejection, loneliness and lack of acceptance.

The second type of general trust relates to an individual (or small group). Bob *generally trusts Alice in particular* if he believes her to be generally trustworthy (as in the definition of Chapter 3), that she is well-disposed enough towards him, and so on. He generally mistrusts her if he believes her to be generally untrustworthy. Bob generally trusts a group if he believes its members to be generally trustworthy.

General trust/mistrust is a *chronic* or *persistent* condition, enduring through time, probably robust against the occasional let-down, as opposed to more specific trust as an *acute* condition, of recent onset, in many cases of short duration, fixed to a particular commitment and the specific contexts in which Alice and Bob find themselves. General trust may emerge from a series of successful particular episodes of trust.

Generalised Trust and Related Terms

General trust in the above sense is perfectly coherent and evident. However, many commentators have attempted to construct an ideal of *generalised trust*, in which a surprising amount of credence is given, not to the actual judgments and behaviour of individuals, but to their answers to surveys (Seligman 1997; Uslaner 2002; Timothy 2020, 152–154). I shall use "general trust" to refer to the uncontroversial type of the previous subsection, but in other respects I will follow other authors' terminology as I will explain. Note also that some of these terms are used ambiguously in the literature between an attitude and an action.

The key terms are as follows.

- **Generalised trust** is the belief that most people can be trusted. "A sense of optimism and control over the world is more important than experience for generalized trust" (Uslaner 2002, 21).
- **Moralistic trust** is based neither on experience nor cultural similarities, nor even on the default assumption that others are trustworthy. It is a Kantian application of the categorical imperative, to treat everyone, of whatever religion, ethnicity, politics or gender, as if they shared one's values and radiated good will (Uslaner 2002, 17–18).
- **Altruistic trust** "runs the spectrum from acting as if one trusted more than one's actual predictions warrant, through giving the other the benefit of the doubt in translating one's belief into action, all the way to coming to believe predictions about others that are more optimistic than realism would warrant" (Mansbridge 1999, 291). In other words, altruistic trust is an action, not an attitude, but specifically involves placed trust.
- **Strategic trust** "is all about reducing transaction costs by gaining additional information, be it positive or negative. … Strategic trust reflects our expectations about how people *will* behave. Otherwise there is no deal" (Uslaner 2002, 22–23, his emphasis).
- **Predictive trust** is the probabilistic reckoning of the likelihood of betrayal by the trustee (Mansbridge 1999, 292–293). It is contrasted with altruistic trust, so is presumably meant to mean acting in exact accordance with one's actual predictions, not giving trustees the benefit of doubt.
- **Particularised trust** is biased toward one's own kind. "Particularized trusters have positive views of their own in-group and negative attitudes towards groups to which they do not belong" (Uslaner 2002, 28).

Those who posit trust in these terms argue that only trust that is morally driven or altruistic reflects well on the trustor. Generalised trust is based on moralistic trust (Uslaner 2002, 21), and distinct from it (Uslaner 2002, 26n.16). Altruistic trust isn't quite the same as moralistic trust, as it is related to predictive trust as well – it gives the trustee the benefit of every doubt (and then some) in the predictions, and, as we have seen, this type of trust is quite common in certain relationships. Yet it is, as defined, a type of evidence-based judgment, whereas moralistic and generalised trust are imperatives to treat all people the same. The difference between those two is that generalised trust can waver, moralistic trust doesn't. "Sometimes things look good, sometimes they don't. Our values (moralistic trust) don't change readily. But the way we interpret them does reflect some experiences from daily life. … Generalized trust goes up and down, though it is basically stable" (Uslaner 2002, 26).

These kinds of trust are stipulated to be ethical improvements on strategic, predictive and particularised trust, which make more of a point of the trustor's self-interest. Particularised trust may be seen as a kind of prejudice, although knowledge of one's own in-group and the coincidence of interests means one's judgments will be more accurate. Strategic and predictive trust are similar; if there is a difference, it is that strategic trust invokes a search for relevant information, while predictive trust is a probabilistic calculation. If that is correct, strategic trust is more open to development and update.

All of these are kinds of trust as I defined it earlier in the chapter, characterised in terms of their warrants. It is noticeable that they only cover a small number of warrants – normative demands, reputation, knowledge of trustee's behaviour, membership of in-group. Hence they exclude trust based on alternative warrants. They may appear to be an exhaustive set, but they are not.

I will roughly group the first three together under the term "generalised trust;" it will be the topic of discussion on occasion through the book. I will sometimes refer to "strategic trust" as roughly opposite to generalised trust, grouping the last three together as types of trust in which the trustor takes his self-interest seriously. It should be plain that generalised trust is a different kettle of fish entirely from general trust, as defined in the previous subsection.

Some commentators, such as Seligman and Eric Uslaner, deny that (what I call) trust is *really* trust at all, because for them trust is a moral imperative. Trust, on this generalised and decontextualised view, demands radical uncertainty about and alienation from the trustee. It is no great feat to trust people one knows well or understands. It is not even very much to trust a stranger, if she is a known quantity in looks, culture, language, etc. For trust to be socially and ethically valuable, one must be disposed to trust wholly unfamiliar persons.

Uslaner claimed it was an example of the golden rule – we should trust others because we should treat them as we want ourselves to be treated (2002,

17–18). Seligman argued that in the modern world, trust must be of "some unknown, unverifiable aspect of alter's behavior," and that "it is in the very otherness of the alter that one puts one's 'faith' and not in any communality of traits shared" (1997, 45). Trusting similar or related people is different in kind (particularised) from real, generalised trust. Serious trust is "unconditioned by normative role expectations" and also "unconditioned by any idea of a generic identity between the interlocutors … [It is] predicated on their fundamental difference, a difference that makes the will, intentions, and calculations of the one opaque to the other" (Seligman 1997, 55). For Seligman, generalised trust applied to everybody unconditionally is trust *simpliciter*, while trust based on affinity is mere confidence. The latter has no merit attached to it, is transactional, aimed at gaining advantage, and illicitly discriminatory against outsiders.

Generalised trust is *not* a basic assumption that others are trustworthy, but rather an imperative to treat all people *as if* they are trustworthy (Uslaner 2002, 17–19). As we have seen, some roles, such as lawyer, priest or parent, do have that kind of imperative attached to them (but even then only in relation to a fixed audience, in limited contexts), but it is surprising to demand such an indiscriminate view of humankind from individuals in general, independent of the roles they occupy.

… And the Survey Said …

If generalised trusting is a different species, then observing actual behaviour won't help to identify it. One's ability, capacity or propensity to trust strangers is unconnected to the many activities and collaborations with people who are similar or known to oneself, because "for most all types of both formal and informal social contact, trust is *neither a cause nor an effect*" (Uslaner 2002, 135, his emphasis, *contra* Putnam 2000).

How, then, do we find it? It is usually evidenced by a positive answer in a survey to a question such as: "generally speaking, do you believe most people can be trusted or can't you be too careful in dealing with people?" (Rosenberg 1956). Is there a point to this standard question? As an indicator of psychological characteristics, it seems hopeless. To begin with, we quickly reach a regress: can we trust those who answer "yes?"

Do respondents give what they believe to be the socially "right" answer? Is it even meaningful – without a context, isn't it a bit like asking "generally speaking, is food tasty?" It is completely unmoored from any actual instances of trusting anyone (O'Neill 2002, 9–10). Even if, as in some surveys, some indicative background is given, the respondent is not actually in that situation. "We should not equate a person's moral stance with her intellectual version of it, nor suppose that a person necessarily knows the relative strength of her own motives and emotions" (Baier 1994c, 66). Trust is a tango between our personal and social selves, a "spontaneous relationship between self-repression and individual interest" (Urbinati 2015, 63).

No doubt statistics about large populations are of some interest to some people. It may be that the number of people who believe most people can be trusted went up from x_1% to x_2% in the three years since we last did the survey, or y% of women believe most people can be trusted compared to z% of men. I suppress a yawn, but maybe it is interesting to some. But what it means, say, to be 12% more likely to trust isn't really spelled out, unless you are 12% more likely to blurt out "yes" when badgered by a Buttinsky with a clipboard (for an extended defence of survey data, Uslaner 2002, 51–75, especially 68–74).

Using such statistics, Uslaner built a model based on survey evidence from mid-to-late 20th century America, and uncovered many correlations (2002, 76–114). However, not all convince. For instance, generalised trust correlates with both volunteering and giving to charity, as one would expect. But volunteering time and effort on a worthy cause would seem to demand *more* commitment than merely filling in credit card details on an online form, and therefore the former is intuitively more important for civil society. Yet Uslaner found a *stronger* correlation between generalised trust and giving to charity, than with volunteering. He proposed a number of explanations (Uslaner 2002, 136), but try this one for size: the sort of person who finds it easy to demonstrate his trusting nature by ticking a box on a survey form might also gravitate toward easier ways to contribute to society.

Many other similar statements melt into hot air under scrutiny. Young people who have trusting parents, can influence family decisions, can choose their own friends and feel free to disagree with their parents have a 76.7% probability of being a generalised trustor, whereas those with mistrusting parents with whom they never disagree, and little family influence only score 42.8% (Uslaner 2002, 163). Perhaps this simply means that ghastly precocious spoiled brats have a better sense of how to fill in forms to please social scientists than those precious few who are seen but not heard.

The problem is the free-floating nature of generalised trust, and it turns up again and again. The real world will force its way in. The picture of a generalised trustor, "white, older, and well-educated … secure in themselves and their neighborhoods and [who] had parents who encouraged them to place their faith in strangers" carries a whiff of complacency, whereas the particularised trustor, "more likely to be black, younger, with less education, fearful for their safety, loners without social support systems, and the children of parents who worried about other people" (Uslaner 2002, 111), perhaps might have some justification for being wary. Uslaner may in fact have discovered the new social cleavage between globally mobile citizens of nowhere stuffed full of social capital and those rooted in particular communities (Goodhart 2017).

Generalised Trust as the Rejection of Pluralism

For Uslaner, particularised trust cannot be the "backbone of a civil society," while Seligman criticised Durkheim for citing "intermediary groups as providing that familiarity necessary for the establishment of modern forms of solidarity" (1997, 95). The reasoning is odd. Particularised trustors participate in groups and have friends to the same extent as generalised trustors (Uslaner 2002, 140–141), so associations and friendships can't distinguish between good, generalised trust and bad, transactional, exclusive particularised trust. Indeed, involvement in church and children's groups even *decreases* generalised trust (Uslaner 2002, 134–135), because it familiarises people with similar folk, and raises barriers against outsiders (it "others" them, in the jargon). "Freedom of the other and the risk it implies are fundamental aspects of modern social relations and cannot be 'defined away' through membership in intermediary associations" (Seligman 1997, 95). Nazi Germany was replete with *bunds* – not an example to follow.

But common experiences in civil society and interpersonal familiarity aren't in *tension* with trust. *Only* on the curious definition of generalised trust can they be seen as bad things. The serious problem with generalised trust, even if it captures something other than dispositions to answer survey questions in a particular way (which I doubt), is that there seems to be no reason why it would align with trustworthiness. Neither Seligman's nor Uslaner's account mentions trustworthiness very much at all. If generalised trust is to be extended to anyone, including strangers and those with whom the trustor has little in common, it cannot benefit from a rich system of signalling trustworthiness, because that sort of communication requires links between trustor and trustee, and the deeper the communication, the stronger the links, and the better the alignment. Hence the generalised trustor deliberately limits the amount of input he solicits to support his trust judgments, and so, all things being equal, will be less reliable and more prone to disappointment than the particularised trustor, who at least knows what he is talking about. The net result will be trust placed not only without discrimination, but also indiscriminately.

This doesn't seem satisfactory as an account. "It takes generalized trust to give either time or money to help people you don't know and likely will never see. Good works toward your own kind … [are] based upon personal experiences and reciprocity. Good works toward people you can't identify depends [*sic*] upon your moral sense, and not upon your life history" (Uslaner 2002, 143). When we consider this, we should remind ourselves that trust is not a logical requirement of charitable work, and that in any case we are dealing with percentages and probabilities. The most we can say is that there is potentially some kind of correlation between respondents' survey answers and the good works they do, which does not mean that those who are warier of strangers and favour people like themselves never give to charity. If we want to support the spread of good works – and why should we not? – then why not support the associations and little platoons in which particularised trust thrives?

Indeed, if surveys are all we have, they can provide no concrete evidence that generalised trustors actually give more. This requires empirical work, *seeing* who trusts across different contexts, and who is more circumspect. Even an individual whose self-image is that of a trusting person needs to observe their own judgments and actions in specific circumstances to confirm it. It's one thing to sit in one's armchair thinking benignly of all humankind, and quite another to put one's hand in one's wallet and lend someone £100. Someone who says things like "my problem is that I am too trusting" is usually complaining about being hard done by, not revealing a trusting disposition.

If generalised trust is different in kind from self-interested strategic trust, can we even find it? If it is not inferrable from observable instances of strategic trust, it may not even exist, except in surveys. Seligman thought that trust, properly understood, only applies to strangers about whom we know nothing. He was prepared to accept the unintuitive consequence that "in its pristine form, [trust] is very rare, for we are rarely in the company of others among whom no attributes of familiarity can be imputed" (Seligman 1997, 70), a strange admission that ought to set alarm bells jangling.

We have a paradox of dissonance. Trust is most valuable where there is dissimilarity or heterogeneity between trustor(s) and trustee(s), so that those who argue for the importance of generalised trust suggest it is a kind of duty to help overcome such gaps (Uslaner 2002, 2–3). However, whereas this view of trust wants to force it into a primarily moral template, it will always include expectations that are liable to be disappointed on occasion when diverse people are brought together. The generalised trustor may need to push back if he is to reach a *modus vivendi* with the trustee. If trust is unconditional, the trustee can never learn what is expected of her.

Pluralism is a problem for generalised trust that cannot be wished away. Uslaner assumed that "generalized trusters have a distinctive view of civil society: they see it as *one* society united by a set of common values" (2002, 191, his emphasis). They are "tolerant of people who are different from themselves" (2002, 192). Yet there is an obvious contradiction here: if generalised trustors see society as united by common values, then they must believe that those who are different share those values. It is not that they actually trust heterogeneous strangers, but rather they assume strangers are really like themselves deep down. The difference between the generalised and the strategic or particularised trustor, then, is not that the former is morally superior to the latter, but rather that the former wishes difference away, forcing the trustee into a template, recalling Hayek's definition of coercion quoted in Chapter 2.

Generalised trust, it turns out, is not so different in principle from trust grounded on fellow-feeling or self-interest; the main difference is that the generalised trustor illicitly assumes that everyone is just like him. Without critical engagement with the other, the generalised trustor is likely not only to be disappointed, but also to label the other unjustly as untrustworthy if she fails to come up to standards he has unilaterally set.

Generalised trust is a cosmopolitan idealisation, glorifying non-discrimination without finding any evidence for its value; *general* trust, as I defined it earlier, is a positive character trait, evident in the world, to be encouraged alongside general trustworthiness.

PART II

Conservatism and its affinity for trust

Introduction

Part I's account of trust will now be situated in a political context. Part II will argue that the ideology of conservatism has a strong, possibly unique affinity with trust. Social trust gives a foundation to conservative thinking and policy, while conservatism itself underpins the practices and institutions that facilitate trust in society. Trust is a response to uncertainty and free will that thrives in conditions of stability, order, predictability and managed expectations. A conservative society is therefore optimal for trust, all things being equal. Conversely, those ideologies expressly designed to challenge order and stability impact trust negatively, the topic of Part III.

By "conservatism," I do not mean any of the ideologies, some of them wild and berserk, that have emerged from the fringes of the political right, particularly those with libertarian or narrow nationalist overtones. I refer to *small-c conservatism* (O'Hara 2011), which is not very well represented in party politics as the Puritans and Peasants run wild. It is endangered in its traditional political ecosystems.

Small-c conservatives are obviously antithetical to the Puritans, but the Peasant style of politics (or perhaps paranoid style – Hofstadter 2008a) is being absorbed into traditionally conservative parties. In the Anglosphere at the time of writing, the British Conservative Party is going through a collective nervous breakdown, and hasn't had a noticeably conservative leader since Theresa May (who wasn't conspicuously successful), while the American Republican Party has left its senses. Neither, in their headbanging radicalism, can properly be called conservative.

Can we separate the politics of the Peasantry from small-c conservatism? I think we can. The theory of the conspiratorial Peasants has been encapsulated succinctly in a triad (Barkun 2013):

- Nothing is an accident
- Nothing is as it seems
- Everything is linked

This is a politics of mistrust. For the conservative, it is different. She would like to say that:

- Everything is an accident
- Everything is as it seems
- Everything is linked

It may be naïve, but it's true, and leaves plenty of room for trusting opponents' motives (while doubting their self-assessed competence).

In Chapter 5 I shall briefly characterise small-c conservatism's scepticism of grand plans, radical ideologies and hubris. Chapters 6, 7 and 8 look at how conservatism enables communication, community, and the state, via its support for trusting relationships.

5. Small-c conservatism

Conservatism and trust go hand in hand. Conservatism is a Protean ideology, because what one wants to conserve, and the threats one perceives, will vary through time and space (Huntington 1957). Nevertheless, "something has remained the same, namely the conviction that good things are more easily destroyed than created" (Scruton 2017b, 127). Both conservatism as a philosophy (O'Hara 2011) and trust as a social practice (Luhmann 2017) are concerned with risk, an ineradicable aspect of innovation and change. Even in a highly dynamic and innovative society, stable foundations are a *desideratum*. Innovation that threatens the foundations will erode existing trust by removing fixed points of reference, a point which underlies the philosophy of *contextual integrity* (Nissenbaum 2010), discussed in Chapter 6.

For instance, fiscal orthodoxy and sound money are usually valued by conservatives, because they enable trust in economic life. Inflation erodes the reliability of money, whose signals become changeable and whose valuations come to be seen as unfair through time (Stantcheva 2024), undermining fruits of prudence and hard work such as savings and pensions. Debts are inflated away, resulting in transfers from savers to borrowers. Excessive public borrowing forces future generations to subsidise the consumers of the present.

RESILIENCE AND RISK

"In the course of a lifetime customs, places, networks, institutions, shared ways of being all amplify our attachments, and create the sense that we are at home in the world, among familiar and trustworthy things. That sense of the familiar and trustworthy is precious to us, and its loss is an occasion for anxiety and mourning" (Scruton 2017b, 12). Yet the vulnerability of those shared ways of being is remarkably underemphasised in political and social thinking. Many political ideologies blithely assume that ideologically-driven social change won't undermine existing valuable institutions, so that we can both carry on as we are, and create visionary futures.

The world is not naturally stable in the medium term. Happenstance cannot be eliminated. Not all risks are predictable or calculable, and the major ones tend to be the hardest to model. However well-entrenched a community may seem to be, it can prove fragile in the event of sudden upheaval, when familiar orienting devices disappear. Every decade or so something happens to disrupt global systems, from the attack on the World Trade Center, to the financial crisis, the Arab Spring, COVID, the Russian invasion of Ukraine, and so on and so on, all against a background of increasing geopolitical tension, climate change and the radical technological flux of digital modernity.

For conservatism to be a positive programme, rather than simply reactive, policy needs to support the resilience of communities. Those places where austerity was implemented after the financial crisis of 2007–09 (often criticised by economists – Blyth 2013; Schui 2014; Fender 2020) benefited when the COVID pandemic struck, because sound public finances laid the groundwork for fiscal stimuli. Those nations with high levels of debt were less able to support damaged economies. After the pandemic, those nations which splurged on stimulus found themselves in a trickier fiscal position, endangering future resilience.

Conservatism is not only about major shocks. It makes a wider claim that any functioning society is so complex and dynamic it cannot be managed or predicted as a system, for three reasons. First, it is a massive and highly connected *network*, so that events that directly affect some parts can ripple through to other areas indirectly with unpredictable and unintended consequences. Second, many of the social phenomena with which we are most concerned are *emergent* from complex underlying mechanisms upon which they supervene – an inflation rate (for instance) emerges from billions of small exchanges in a particular currency through time. Emergence means that it is impossible to legislate directly at the macro level to regulate the emergent phenomenon; instead, a policy is needed that will affect the micro-level phenomena upon which it supervenes, a far harder task that may not result in the desired macro-level effects. Furthermore, a third problem is that once policy is in place, people adapt to it. Such *reflexivity* means that a model developed from data describing the *status quo ante* may fail to predict the sequel; for instance, regulations designed to increase safety may result in riskier behaviour from a reassured population.

Policy directed at a particular problem therefore encounters three significant difficulties. First, the models upon which it is based are simplified renderings of the immediate context to make the problem tractable (Scott 1998, 134, 264, 288ff.). Second, the models cannot include the wider context of the whole social fabric or network, and so cannot predict the indirect consequences elsewhere. Third, when the policy is implemented, the models are already out of date, because the data upon which they are based does not include the reflexive post-implementation response.

The conservative postulate is that the mere existence of particular laws, institutions and practices through time is (defeasible) evidence that they play a useful role, even if it is not evident. Furthermore, people will have adapted their behaviour and future plans to them. The upshot of conservatism is that resilience rests on well-understood procedures, preserving the capability to react to events small and large that are in essence unpredictable. If things are going wrong, change is essential, but needs to be incremental, rigorously evaluated and reversible where possible to manage risk effectively. For that to happen, it needs to be implemented by sympathetic institutions without a wider agenda for social change imposed from the top down.

A QUESTION OF DEFINITION

"Conservatism" has been tricky to define, or at least for political philosophers to agree on a definition. I propose to take seriously the dictionary definition of "conservative" with a small "c," as a concern with, even aversion to, change. One would think this was straightforward, but the word is so badly misused that it is worth spending a page or two dismissing a common and pernicious misunderstanding.

"Conservatism" in American Political Rhetoric

The word "conservatism" is *not* intended to recall the indiscriminate, and regrettable, use of the term in American political discourse to mean anyone to the right of politics, whatever their principles and policies, lumping together those who want a big state (George Bush Jr), those who want a small state (Rand Paul), defence hawks (John McCain), pro-business types (Mitt Romney), religious and cultural fundamentalists (Pat Buchanan), tax hawks (Grover Norquist), as well as people with no discernible principles at all (the 45th and 47th President of the United States). The term "liberal" is equally misused in America, to mean anyone on the left of politics, conflating it with socialism.

These erroneous Americanisms have driven much academic research about the differences between self-described "conservatives" and "liberals" – self-description rendering the research almost useless. It also encourages outlandish and unjustified research methods. For instance, Uslaner, in constructing his trust model, needed a proxy to measure the selfishness or materialism of the American people (naturally in a survey no-one would answer "yes" to the question "Are you somewhat/very selfish?").

> Alas, there is no measure of materialistic values over time. A surrogate is Stimson's … measure of public mood. The mood construct is an estimate of the relative liberalism (higher values) and conservatism of the American public across a wide range of policy issues over time. My logic for using it is rather simple: Materialism rises in

> more conservative eras (such as during the Reagan administration), which are often called by such names as the "me" generation. (Uslaner 2002, 188)

This is extraordinary. A measure of selfish materialism is required. A measure of public mood on a "liberal-conservative" (i.e. left-right) scale is used instead, because people are supposedly more selfish when the mood is conservative (no evidence given for this). So it is deduced that trust declines because materialism rises (the correlation asserted without evidence) during conservative periods! Well, there's a surprise. Hidden away in the footnotes, it turns out that those who claim to trust are slightly more likely to identify as conservative than liberal (Uslaner 2002, 195n.6). However, neither claims from surveys nor ideological self-identifications are worth very much, as I have argued.

Worse, as US politics has become more polarised, even the bastardised use has warped, and "conservative" has come to connote ever more extreme positions. A "conservative" is now supposed to oppose immigration even when it is legal, deny the human contribution to climate change, insist on the right to tote automatic weapons, refuse to admit the legitimacy of federal regulation, and oppose taxation on principle. It is hardly the responsible and sensitive ideology of Burke and Madison, who would now be called CINOs (conservatives in name only).

A Working Definition of Conservatism

Pace fevered Americans, Conservatism is a type of scepticism about policymaking and the world (Kekes 1998, 30–31; O'Hara 2011, 23–49). While it is often associated with particular policies, practices and institutions, small-c conservatism can be boiled down to a pair of principles. First, the *knowledge principle* (O'Hara 2011, 49–51) suggests that the connected, reflexive and emergent world defies the attempts of policymakers to engineer new opportunities or right perceived wrongs. Furthermore, they cannot tell for sure what benefits would be foregone were a particular institution altered or removed.

This by itself won't give us conservatism,[1] so a second principle is needed: the *change principle* (O'Hara 2011, 52–90). Given the knowledge principle, there is always a risk that policy innovation will fail to achieve its aims, and mess up positive institutions. No social science, quantitative or qualitative, can prove these risks are negligible. Consequently, political change ought ideally to be incremental, rigorously evaluated and reversible where possible.

---

[1] Some have unaccountably read me as suggesting that the change principle can be deduced from the knowledge principle, which it clearly cannot (Clarke 2017, 552n.2; cf. O'Hara 2011, 51).

Does that mean nothing gets done? No, it simply means that the risk of change should be factored into cost-benefit analyses. Conservatism is sometimes erroneously stated to be a bias toward the *status quo*. This is false: rather, conservatism tries to counteract the bias of "progressive" or rationalist ideologies *against* the *status quo* (O'Hara 2021, 82–86). New policies are certainly not banned, but the burden of proof that they are likely to be beneficial rests with their supporters, and the meaning of "beneficial" should not be left to them to decide.

The size of the risk incurred by change is inversely proportional to the positive qualities of the society in question. If a society is unpleasant, then the risk of change is relatively small; if it is pleasant, then it is proportionately greater (Kekes 1998, 9–12; O'Hara 2011, 97–106). "To make us love our country, our country ought to be lovely" (Burke 1999a, 172). A particularly serious danger sign relevant to our inquiry is when trustworthiness is in short supply in a society, or when it is rationed to kinsfolk or other close associates.

CONSERVATISM, THE PRIVATE SPHERE AND CUSTOM

Conservatism relies on trust – of one's fellows in communities and associations, of the wisdom of one's ancestors whose customs and traditions guide (but do not prescribe) today's conduct, and of authority. As Burke had it, each person's private stock of reason is relatively small (1999a, 182), and these scaffolding reinforcements are essential for navigating a complex public world, providing a framework for the mutual benefit of those within it.

Hume's moral anthropology emphasised the family, community, reciprocity, collaboration, obligation and sympathy – all means by which trust spreads and is sustained. Baier's commentaries on Hume and critique of Kant's "masculine" moral philosophy emphasised the importance of trust-sustaining structures across society, not only public institutions supporting the liberty of free, equal and autonomous persons, but also more private worlds concerned with unchosen, rich and intimate relations, less abstract worlds of care and self-sacrifice (Baier 1994c, 1994d). One needn't endorse Baier's own brand of feminism laced with *ressentiment* to appreciate her profound extension of moral theory into areas of great interest for conservatives, especially the family, childcare and small associations, without which more exalted public institutions would be hollowed out, and individuals would struggle to retain their individuality.

The private and public spheres need to be in balance. If the individual retreats into her own private affairs, she will lose influence on the conditions of trustworthy public life, ceding it to elites (de Tocqueville 2003). Demand for individual freedom and retreat into the private sphere will tend to undermine social norms, rendering them less effective in the promotion of general trustworthiness. All too often, this results in the expansion of the state, which has to step in to remedy resulting problems (Deneen 2019, 17).

The resources of the private sphere need to be available to public life, not separated from it. The social background was important for Hume, creating the stability from which trust, as well as knowledge and the useful arts, can grow: "a [free] republic … gives rise to LAW, even before mankind have made any considerable advances in the other sciences. From law arises security: From security curiosity: And from curiosity knowledge" (Hume 1987a, 118). An unfree society, excess of authority, blind deference or "tyrannical usurpers over human reason" (Hume 1987a, 123), on the other hand, suppress learning, trust and understanding.

Custom and habit enable us to engage with the world and each other. Enlightenment philosophers demanded that they be justifiable by reason, or eliminated. On the contrary, Hume would justify these informal epistemic guides, if they stood in need of justification, in terms of their reliability and practicality (Hume 1978, 12–13). Some ways of thinking may be rejected because they are "neither unavoidable to mankind, nor necessary, or so much as useful in the conduct of life" (1978, 225), but some habits will withstand a deep survey by a range of people using what Hume called their understanding, including memory, experiment, logic, causal inference and so on (1978, 620–621). That is not to privilege a particular source of epistemic authority, but rather the aggregation of a range of authorities of varying credibility and trustworthiness (cf. Wittgenstein 1979, §603). When Hume did evoke reason, he was typically attempting to widen its definition from the desiccated rationalist understanding, to include "our capacity to coordinate our speaking and our actions with the speech and action of our fellows, to coordinate moral and aesthetic judgments as well as factual and mathematical ones" (Baier 1994d, 94).

What standards do we use to determine what is "so much as useful in the conduct of life"? Obviously, it would be those of our culture and community (O'Hara 2022b). Why would a conservative go to such great lengths to defend and preserve her culture if she didn't then want to use its ethical and social resources in her political thinking? Whose values would she use otherwise? Why should she prefer their values to her own? And within a community, aren't its values those that its members can agree on during reflective moments, and when reacting to new phenomena, whether welcome or unwelcome?

Isn't this circular, judging our own society by our own standards? Of course it is. So what?

SMALL-C CONSERVATISM AND POLITICAL TRUST

Like Onora O'Neill, I am sceptical about there being a crisis of trust in the sense of justified disgust at public untrustworthiness, and am drawn instead to her diagnosis of a culture of suspicion, characterised by the performative expression of mistrust (O'Neill 2002, 44–45). Political actors (broadly drawn as in the Introduction) can be sorted into two groups: those who have an interest in preserving and improving the current pattern of living, and those who would like to tear it down or disrupt it. Those in the second group (Peasants, Puritans, populists, nationalists, eco-warriors, and radicals of all stripes) have much to gain from a culture of suspicion, and Part II is not concerned with them. Conservatives are proud members of the first group, concerned as they are to appreciate the goods of society, and keep them as far as possible in trust for future generations. They are not the only members of the first group, however – many socialists, liberals, environmentalists and others share this aim too.

It behoves a politician of the first group to try to promote trust not only in politics, but also in institutional processes in general, as these facilitate the smooth functioning of society as a whole. One obvious way of promoting trust is by being trustworthy, which is within the politician's power. However, it is not sufficient, since political trust is the *belief* (of citizens) in their trustworthiness, a complex set of attitudes as we know.

An equally obvious way of promoting political trust is to show civility to one's fellow political actors, even opponents, as long as they share membership of that first group. Keeping criticism constructive and avoiding hyperbole will enable actual cooperation when necessary, and a general move back to getting-things-done politics. It follows that the polarisation of modern politics is a disaster for the resilience of society, because beggar-thy-neighbour politics solves nothing. All politics is about difficult choices, but if the policies of one government are immediately undone by their successors, there is no possibility of long-term planning.

A resilient society demands non-partisan thinking and solutions that can outlive the electoral cycle. High-trust communities with strong social coherence are needed for long-term planning. In the Netherlands, disastrous floods in 1953 prompted the complex and comprehensive coastal defences that took over 30 years and cost the equivalent of €5bn to build – an engineering masterpiece, but also an extraordinary *political* achievement (De Mulder 2019, 19–20). Such achievements need to be more routine.

For example, in the UK, managing the green transition of the economy, upgrading infrastructure for energy, water, telecommunications and transport, uniting health and social care into a single system, maintaining defence and cybersecurity in a world of many enemies and expensive technology, restoring the criminal justice system, determining its post-Brexit relations with the EU and the rest of the world, managing immigration and asylum, not to mention rebuilding Ukraine, will be expensive and require programmes that will outlive individual governments. That is not to say that consensus between opposed parties on the detail is either likely or desirable, but rather that without broad agreement on the nature of the problems and the general direction of the solutions, no progress will be possible in the medium to long term.

In that context, listening to the moans of opponents is a good early warning system for problems – at least some of which will be genuine. High-bandwidth, good faith communication across the aisle naturally helps understanding, and is therefore no bad thing for trust (especially where mistrust is common). Constructive engagement with opponents' agendas constitutes what I have called the *vaccination strategy*: a few minor changes and reforms implemented by a small-c conservative government may do a lot of good, without risking the wholesale change that non-conservatives prefer (O'Hara 2005, 136–138).

More ambitious reform of political structures and procedures is unwise, unless they plainly aren't working. The unintended consequences can be extensive. Initiatives to change British Parliamentary procedure, for instance, have had largely negative effects. Family-friendly policies have diminished the amount of time that Parliament sits, from 173 days at an average of 9 hours per day in 1988, to 147 days at 7.5 hours per day in 2021 (Carthew 2023), while the amount of legislation to scrutinise has risen inexorably. Worse, by limiting evening sittings they have reduced the likelihood of rival MPs mixing socially. These efforts were introduced to increase the diversity of Parliament, and were indeed successful, but they correlate with its decline in public standing and the rise of beggar-thy-neighbour politics. The reason for this, I suspect, is that diverse members have less in common and prefer to remain within their cliques. Postal voting and other novelties have made it easier to vote, thereby probably increasing turnout, but they haven't restored and can't restore the trust in politics whose loss was the cause of low turnout in the first place. All they have done is made it a little likelier that someone who doesn't trust politicians votes, alleviating the symptom but not the disease.

I am sceptical of the contribution to political trust of firm commitments, costed promises and transparency, which have also proliferated since the waffly, not to say meaningless, manifestos of the 1950s. Since the background context may alter between promise and policy delivery, too firm a commitment may saddle the politician with policies that are now irrelevant but politically costly to drop. It's great to promise tax cuts, but if a massive increase in government spending is needed (a pandemic, a war, a banking crisis anyone?), then the promise either will have to be broken, or, if honoured, will make things even more difficult. Specific promises are useless – much more important is to cultivate a reputation for competent, constructive, flexible and non-ideologically hidebound government shoring up resilience (a reputation that can be burnished in opposition).

Being trusted involves sending the right signals to would-be trustors *consistently*. Voters' memories may be short, but they are not non-existent; if today's signals contradict yesterday's, there is no reason to think that voters will concentrate only on the new messages. If voters felt excluded from a politician's audience, then getting them to believe they are now included is a matter of extended persuasion, not mere assertion.

Political trust has to do with reliability, adherence to principles, framing and pursuing common interests, and a certain predictability grounded in good faith. It involves making and framing commitments that voters and citizens can connect with – a Communist might well be utterly trustworthy, but nobody wants the things that she commits to. It involves seeking and understanding the lives of citizens, to appreciate the good things and protect them from incursion. Recalling Burke at Bristol, this doesn't mean the politician has to be deferential to her voters – indeed, that can't be possible given our heterogeneous electorate – but she does need to be sensitive to their values and perceptions. She also needs to see wider (national) interests, and be able to extrapolate from policies to their likely (and unlikely) effects at local level (an inexact science). Trust is essential for the smooth execution of authority; without a positive belief in the motivation, abilities and incentives of those in authority, whether they be politicians, religious leaders, experts or financiers, citizens are unlikely to submit without protest to their advice.

Hence the representative (whether or not conservative) needs to adhere to the Burkean principle of responsible judgment to signal her trustworthiness and foster trust. She also needs to make commitments that place voters in her intended audience of beneficiaries. This is less to do with specific policies and eye-catching schemes, than a general sympathy for the ways in which people manage their work, leisure and security. It is less to do with supporting unorthodox experiments in living and protecting the interests of the antisocial and criminal, than with protecting humane social norms. Such is the connection between trust and small-c conservative politics.

6. Legibility and contextual integrity

Chapter 5 sketched the conservative ideology's dependence on, and support for, trust. In this chapter, I will suggest a ground for the strong connection. Part I showed that trust and trustworthiness depend on an understanding between trustor and trustee, in particular their agreement on (i) the content of the trustee's commitment, (ii) its application in the trustor's context, (iii) the trustor's membership of the trustee's audience of beneficiaries, and (iv) what the commitment entails in the context.

If some social institution, process or practice has been around for a long time, it must fit, and have fitted in the past, into many practical contexts, and served the various purposes of many different people. It is therefore more likely to be socially useful and valuable. That's not always true (the Mafia has been around a long time), but it's a good indicator. The roles we adopt within or with respect to such institutions and practices are negotiable to a degree, but only within constraints, and cannot simply be created by individuals for their own benefit, or to their own design (Seligman 1997, 23–24). They evolve at a leisurely pace in response to social pressures (too quickly for traditionalists, too slowly for radicals).

Some roles liberate, presenting us with a menu of actions, while others constrain, helping us behave socially and acceptably to our fellows (Schoeman 1992, 17). Consistent but evolving roles, institutions and practices also connect us to our past, helping us understand and empathise with our forebears. They are woven into our historically- and geographically-based communities, and help us understand our social world a little better by giving us something in common with similarly-located others.

This understanding is important for what James Scott has called *legibility* (1998). An environment is legible to someone if they can navigate it, feel comfortable in it and familiar with its vagaries, work with it or within it to achieve their goals, and use it to find resources. Many types of knowledge support legibility, ranging from science and economics to myth and ritual, especially when practical and rooted in behaviour (Nyíri 2016). Legibility is, ultimately, *performed* – we *collaborate* in carrying out rituals, reinforcing narratives, referring to symbols and reacting in intelligible and unironic ways. This makes people legible to each other.

Left to their own devices, communities devise ways of ordering, measuring and distributing goods, exploiting local knowledge and the resources to hand, navigating via features of local history and geography, all specific to the context and in their own interest. The result of these local orderings is that the accuracy of community members' mutual trust is enhanced. What looks like chaos to the outsider is intensely meaningful to the community, immanent in its forms of life (Scott 1998, 279, 305), reproducing and disseminating skills and practices (Scott 1998, 301), and specifically facilitating understanding of their fellows' aims and goals. At least in our unsophisticated moments (which is to say, most of the time), we take fidelity to social standards and clichés as a proxy for trustworthiness, because they demonstrate willingness to be understood and read by an audience.

Changes in the environment or these practices can decrease the sense of connection, as they become less legible. I have already mentioned money, trust in whose reliability underpins economic life. Unsound money and high inflation not only sever an important connection between past and present (e.g. past work paying, via savings or inheritance, for today's wellbeing), but also render exchange and employment less legible, so that consumers and workers are less sure of value, less able to judge their own interests. Excess or hyperinflation has disastrous psychological, as well as economic, effects (Salerno 2013; Fajardo & Dantas 2018).

It is hard to ring-fence the environment and the practices around it to protect it jealously from outside influence (the Peasant model), or alternatively to remain utterly open to any kind of influence at all, to refuse/deny ownership of one's own traditions (the Puritan model). These traditions have value in underpinning legibility and trust, and yet should be open to influence to enable the society to develop organically and to be open to trade or migration. There needs to be a middle way, preserving legibility but not at all costs.

Scott critiqued the bureaucratic state for rendering its citizens legible to it, in order to co-opt their powers and resources, e.g. for tax collection or military conscription. It simplifies (or "modernises") their behaviour and contexts with administrative orders, inventories of land, standardised measurements of human resources, codes of deviance, religions and classifications of ethnicity (Scott 1998, 51–52, 77, 183ff.). Such top-down measures work primarily in the bureaucrats' interests, often for purposes remote from those of their objects. The knowledge they presume is not the rooted, local knowledge present in communities, but the sort of abstract theory that falls foul of the conservatives' knowledge principle. Modernisation can certainly provide benefits, but it has its own logic that can be alien to unwilling communities.

LEGIBILITY VERSUS FAMILIARITY

Legibility goes beyond mere familiarity. Luhmann cited familiarity as a precondition for trust and distrust (2017, 22), but it can only manage a restricted set of situations. Legibility is forward-looking and generalisable, whereas familiarity is restricted to experience one has already banked. To illustrate the difference, take two chessboards. On one, place the pieces more or less at random; on the other recreate the position half way through a genuine game. If we show the boards to an experienced player, he will be *familiar* with neither. But the second will be *legible* to him – he will be able to suggest possible moves, understand the strategies of the two players, and estimate their abilities. The former will be meaningless, and he won't know what to say (Chase & Simon 1973).

The environment is legible to Bob if he can read it; he need not be familiar with it. If Alice is familiar to Bob, he is acquainted with her and has had some exposure to her behaviour and reputation. To say that she is *legible* to him implies a deeper understanding that applies even if she is unknown to him. He can read the expressive codes she uses for the signals she sends. If her commitments are implicit and open-ended, he will share her understanding of them.

Luhmann argued that as the social order gets more complex, there is a greater need for trust, and (on his theory) a corresponding need for more familiarity (2017, 23), but complexity makes things *less* rather than *more* familiar. When familiarity fails, it can only come back as system trust (Luhmann 2017, 25), or in my terms it switches from local to global trust. Yet even on Luhmann's reading the global system is not familiar. Global trust requires *understanding* of the systems for verifying the trustworthiness of strangers, not *familiarity* with them. They need to be *legible* to a wide range of trustors, who rely on them occasionally, without the practice and mastery that familiarity would create. Indeed, they may be designed specifically to be legible to inexperienced laypeople.

LEGIBILITY, TRUST AND TRUSTWORTHINESS

The existing order is legible to those within it, and for that reason, helpful in understanding not only how the world works, how to navigate round it, how to adapt to it, and how to change it, but also how our fellows will navigate through it, what their commitments amount to and how they are limited. Legibility also benefits from a history of interaction evident in traditions of institutions and practices, which connects a community to the actions and projects of its forebears, helping establish collective memory (Hume 1978, 554ff.; Burke 1999a, 192–194). The resulting social identity, moulded by order, habits, education, memory and other common resources, then functions as a connector via language, social norms, institutions and practices. This simplifies the processes of assessing trustworthiness, while simultaneously deepening and enriching reciprocal commitments. None of this is true of order imposed on communities from the top down.

Legibility cannot determine whether a trust relationship will be established, but it affects the quality of the interaction and the ability of the trustor to assess the trustworthiness of the trustee. If Alice is legible to Bob, then he is surely better able to determine her trustworthiness. This can be expressed in terms of fast and slow thinking (Kahneman 2011). Where strong cultural connections underpin mutual legibility, building trust can be straightforward, instinctive and emotional (though mistakes might be made, and trustors hoodwinked). However, where people have little in common, working out expectations demands a big cognitive overhead. The conservative concern about modernity and rationalism is that they undo and banish precisely those common experiences which promote legibility and therefore trust, often as a deliberate strategy (as Burke warned about the French Revolution).

One particular means of preserving common experiences is *contextual integrity*, defined by Helen Nissenbaum as the conservation of the important norms of a context as it evolves or is changed (Nissenbaum 2010, 140). Nissenbaum focused on norms about the flow of information, but they could also be moral norms, behavioural norms, conventions and expectations, norms of politeness and etiquette, and so on. When contextual integrity is respected, essential aspects of the context remain intact through time, preserving legibility to its occupants.

It is tempting to apply one's own values to a social norm structure, and to critique "maladaptive" norms, i.e. those of which one does not approve (Bicchieri 2017, 1). However, the point of norms is that they are widely accepted, and one person's maladaptive norm is another's familiar lodestone. If norms are to do their work of promoting cooperation and trust, they have to be *accepted.* If they don't meet elite standards, we still need a better reason for changing them. Far better to have trust and cooperation than theoretical purity (O'Hara 2023, 149–154). That does not mean being uncritical: "being culturally embedded is not the same as ritually and unreflectively mimicking whatever others do. Cultural embeddedness allows more subtlety than this equation suggests. We have no reason whatsoever to think that those who are culturally sensitive are more disposed to morally outrageous conduct than those who are oblivious of cultural norms. Indeed, we have every reason to think otherwise" (Schoeman 1992, 5–6).

If Alice is keen to preserve contextual integrity, that *in and of itself* is a signal of her trustworthiness. Protecting or preserving norms facilitates social cohesion and legibility, so her actions are calibrated not to confuse or mislead others. She is prepared to work within accepted social parameters, and does not insist on imposing her own view of the world upon everyone else – an important signal of respect for others.

Where Alice and Bob embark on a trusting relationship, whether reciprocal or not, they are typically concerned to make themselves legible to each other, for example using conversational norms such as turn-taking and being informative and relevant (Grice 1989). By being civil and polite, using codes of language, dress and behaviour that are respected across a community, Alice makes herself *more* legible to Bob, as well as signalling to him that she respects him and is prepared to adjust her behaviour to meet accepted standards for their mutual benefit.

Manners are a vital background to trust, especially the civic virtues of wit, truthfulness and friendliness (Aristotle 2011, 1126b-1128b): "for Aristotle, the three civic virtues of agreeableness are central moral goods since … they are enabling goods both developmentally and as necessary conditions of cardinal virtues. They provide the framework within which trust is built and becomes operative" (Mooney & Roy 2021, 186). They are important for civil conversation, establishing a person as well-disposed to another, and therefore liable to be trustworthy. Like all Aristotelian virtues, they can be taken to excess (buffoonery, forthrightness and obsequiousness), and require practical wisdom to apply them in the social world. Such wisdom requires relevant experience, facilitated by the legibility of that world (Aristotle 2011, 1141a–1142a).

Rousseau believed politeness to be opposed to authenticity, but Rousseau's endorsement of a principle is usually a good reason to reject it. Liberal education policies that refuse to apply normative standards of politeness, dress, "proper" speech and pronunciation, spelling, handwriting and so on to students (in theory so that deprived students aren't disadvantaged) end up penalising those who are unaware of such standards or who adolescently refuse to adhere to them (for the additional value of good handwriting for cognitive development, Roessingh et al. 2019). They will have problems communicating their benevolence to their fellows.

LEGIBILITY FOR WHOM?

A vital question is "legibility *for whom*?" The focus of conservatives tends to be preserving legibility of existing centres, with histories and geographies that are precious resources for long-term occupants of relatively settled communities (e.g. Weil 1952; Scruton 2017a). However, the nature of legibility will be a matter of political and cultural debate, and there is no guarantee that conservatives will agree even among themselves on what is important.

Some may focus on legal and political arrangements (for instance, Parliamentary sovereignty and connection with a local MP), while others may consider social or economic matters. Moral legibility may be important for those baffled by the unfamiliar complexities introduced by "progressive" thinkers looking for and finding social explanations for criminal behaviour while undermining traditional moral authorities (Kekes 1998, 79–89). Many communities perceive new rules designed to integrate them with a wider world as "unfair," removing the privileges of long-term membership (this may lead to "progressive" criticisms of a "sense of entitlement" – O'Brien & Major 2009). An influx of new residents with values and culture different from longer-term occupants may create legibility pressures which a conservative may wish to address – this need not necessarily mean people of a different ethnocultural or religious background, but could also include "gentrification" (Smith 1996) or tourism (Pinkster & Boterman 2017). There are strong links between law, politics, culture, norms, institutions, economics, morality, justice and technology, and even a conservative concerned about legibility in one area will in all probability find herself having to take account of considerations from elsewhere.

Suffice it to say that legibility will be contested. Conservatives are likelier to focus on the concerns of longer-term populations than those of transient or more recent ones (cf. Roger Scruton's idea of *oikophilia*, "the love of the oikos, which means not only the home but the people contained in it, and the surrounding settlements that endow that home with its personality" – 2017a, 86).

But the concept of legibility might easily be co-opted by non-conservatives. Legibility for incomers, whether they be foreign investors, businesspeople, engineers, administrators, immigrants, tourists, colonists or refugees, might necessitate active change, to enhance trust between the indigenous population and the incomers. To a highly-educated, geographically and socially mobile professional, legibility consists in rational, technocratic modernity; such a person is interested in homogeneity across communities rather than heterogeneity (Marsili & Milanese 2018). Hers is a hegemonic project, imposing legibility to outsiders on communities, and such change, to her, is no threat – she is not a member of the community, and her aim is to establish trusting relations with those within it (for example, to do business), and to connect the community with a wider world (via telecommunications, motorways, global brands, etc.). However, even such a hegemonic incomer may baulk at wholesale change – global elites have to live somewhere, and so are in a sense parasitic on the maintenance of habitable environments by the less mobile (Scott 1999; Scruton 2017a, 83–111).

Indeed, because she is also a member of alternative communities, for example in global metropolises such as London, New York, Tokyo or Singapore, or even perhaps virtual communities, she will feel threatened by nationalism and the retreat of globalisation. When this happens she may uncharacteristically reach for the conservative toolbox to defend the legibility of the *global* habitat (O'Hara 2011, 183–184). In effect, the negative tone of the "Remain" campaign in the UK's Brexit referendum (dubbed "Project Fear" by its opponents) was conservative in its emphasis on the risks of leaving the EU. The strategy failed because it unfortunately neglected to make the corresponding case that Britain's membership had a positive side.

Even within communities, rendering them legible to the state and other outsiders is not automatically perceived as bad, and often welcomed by those oppressed by small-town constraints or local injustices (Scott 1998, 31, 349, 352ff.). Trusting relations across unfamiliar communities are easier as they both become more legible to outsiders, and *ipso facto* each other. The extent to which standardisation is welcomed or resisted is contingent on many things. The local complexities that conservatives aim to preserve or adapt provide value, all things being equal, for people who understand their environments more thoroughly as a result. Whether this value is outweighed by that of becoming legible to outsiders, and therefore better able to trade and receive government services, is a political matter that can't be resolved in the abstract.

Hence the concept of legibility doesn't dictate a particular conservative policy, but it contains within it notions of understandability, continuity and meaning immanent in those institutions, traditions and structures that at least in part have been crafted over time by groups within a community who need to trust to cooperate, resolve conflict, pursue their own projects, and accumulate social and economic capital. The conservative wants to preserve the legibility that contributes to the success of a society, but her preferred notion of narrow, deep legibility within a community will be contested by the champions of a shallower, broader legibility across communities. This need not be all or nothing of course, but the quality of trust that is facilitated will depend on what emerges from the bottom up, what is imposed from the top down, and where the line is drawn.

POPULISM AND LEGIBILITY

Puritans value moral purity over legibility, which puts them directly at odds with conservatives, but the Peasants are more likely to share conservative antipathy to social change. Following a famous, or notorious, speech by Theresa May in 2016 in which she asserted that "if you believe you're a citizen of the world, you're a citizen of nowhere," David Goodhart divided people into "anywheres," the globally networked exam-passing classes, and "somewheres," rooted geographically and comfortable with ascribed identities (Goodhart 2017; Marsili & Milanese 2018). Change (globalisation) has tended to homogenise the world, making it more legible to the "anywheres," while decentring the "somewheres." In surveys, people are more likely to claim that "they do not recognise their country any more" (e.g. Gaston 2018; Young et al 2019). Increased migration makes it harder to understand and empathise with one's neighbours, with whom there is less shared experience. The digital economy has undermined the secure working environment of the "job for life" (Gray 1995), while technology and artificial intelligence threaten to disrupt many trades (Cowen 2013), all disconcerting and worrying for those living through change.

Many, if not all, forms of populism are responses to perceptions that elites are undermining the legibility of our communities, to the detriment of those within them. For example, in 2014, the UK Independence Party's (UKIP's) then leader Nigel Farage criticised unlimited immigration from the EU:

> In scores of our cities and market towns, this country in a short space of time has frankly become unrecognisable. Whether it is the impact on local schools and hospitals, whether it is the fact in many parts of England you don't hear English spoken any more. This is not the kind of community we want to leave to our children and grandchildren. (Quoted in Sparrow 2014)

Pim Fortuyn fused the language of legibility with Western values and human rights:

> Then look at the Netherlands. In what country could an electoral leader of such a large movement as mine be openly homosexual? How wonderful that that's possible. That's something that one can be proud of. And I'd like to keep it that way, thank you very much.[1]

A final example: F.H. Buckley dissected "the anatomy of indifference" that created Donald Trump's opportunity in 2016. He argued that working class Americans feel they are owed respect. Political elites have created a web of arcane regulations that make it difficult for ordinary people to start businesses or undertake collective enterprises (i.e. the world is less legible to them), while the complexity of those regulations provides high-paid employment to lawyers and accountants, to whom this world is highly legible (Buckley 2018). Astonishingly, Kamala Harris' 2024 Vice-Presidential pick Tim Walz was the first non-lawyer on the Democratic Party's ticket since Jimmy Carter, 44 years previously.

Hence legibility is a good that is ignored by the Puritans, while its absence leads to grievances that can be exploited by the Peasants. The populist crisis in the West can be recast as a crisis of legibility.

A perceived reduction in legibility may therefore drive both populism and conservatism. But whereas the conservative's respect for existing institutions is likely to preserve and improve legibility, the Peasants' repudiation of the *status quo* and the elites within it will have the perverse effect of reducing legibility

---

1 https://en.wikipedia.org/wiki/Pim_Fortuyn.

still further. Hence, while conservatives and populists may join forces to critique rationalist or elite policies, conservatism also contains an immanent critique of populism. Peasant governments cannot solve legibility problems they identified while campaigning, because if they sweep away long-standing political institutions legibility can hardly increase. One study of populist governments showed that they dramatically lower GDP per capita over the medium term, increase public indebtedness and undermine institutions (Funke et al. 2023), none of which is a surprise and which in the aggregate reduce the resilience borne of prosperity and legibility.

The Peasants aim to access power by reducing trust in the political class, but they have no mechanisms for building up trust elsewhere, or across society. Their efforts are aimed at securing trust in the charismatic leader. Everyone else is an enemy, conspiring against the people. Their goal is to foster the twin illusions that (a) they have a monopoly of trustworthiness, via their devotion to the leader, and (b) only they trust the people, as long as they follow the leader. As this is based on the pretence that the populists' will and the people's will are necessarily identical (recall Trump's eerie mantra: "I am your voice"), this strategy is doomed to miss, unless "the people" are genuinely homogeneous, with no competing interests. In any even slightly pluralistic society, there will be faction, association and conflict, and trusted institutions will be needed to manage these civilly.

7. Trust and community

Trust has long been associated with community. Mutual trust keeps successful communities knitted together, even defines them, while the strong ties characteristic of community are conducive to trust.

In this chapter I shall explore community in more detail, including the factors that facilitate effective trust. This maps onto our wider project, because members of a community may draw upon the social resources within it to support trust between themselves (social capital). Community is such a valuable resource that it is almost always detrimental to subvert and erode it, and at worst positively damaging, a conservative insight that is all too often neglected.

Many types of people want to intervene or campaign in their communities to improve them, including young people, highly educated people, businesspeople and technologists, socialists, environmentalists, law and order fundamentalists, feminists, free market neoliberals, and many more. We should also add reactionaries and rigid conservatives who resist a community's organic evolution. All of them have respectable points of view, but tend to assume that the problems of a community can be solved by rationalising its institutions and policies, without threatening the current benefits they provide. The Puritans free-ride on the social stability provided by the rest of us when we take community practices seriously; without that structural bedrock, their ludic "experiments in living'" would lead to chaos and confusion. The Peasants, unlike the Puritans, do value community, but so uncritically that they champion the illegitimate and excessive partisanship of insiders against outsiders. Either way, these culture warriors of our times should be resisted. They do our communities no favours.

WHAT IS COMMUNITY?

By "community" I don't intend any special definition: a meaningful social group that is not completely transactional, which maintains some barriers to entry and exit, and that is important and cohesive to its members, who share a common point of connection, perhaps of location, culture, history, institutions, risks, enemies, practices, goals or values. Membership may be voluntary or ascriptive, earned or assigned. It is likely to perdure, possibly changing and evolving. There are more rigorous definitions which may be used if you prefer, but in this brief section I will describe some of the properties I take communities to have.

Communities may have long histories, may monopolise geographical locations, or may have more to do with religious beliefs, interests or cultural pursuits, Burke's little platoons. They may be large or small, continental, national, or located entirely within a city, suburb or parish – even a single building, in the case of a religious community. They may be diasporas (Kenny 2013). In more recent years, some communities have formed almost entirely online (Shadbolt et al. 2019).

A community is generally taken to be relatively sizeable, which means that it is likely to contain many who are strangers to each other. It will usually, if not always, exceed kinship ties. From some perspectives, its members will look fairly homogeneous, and from others, highly heterogeneous. Some communities emphasise their homogeneity, but it may be irrelevant for others. In some communities, members may be surprised to be told they have anything in common at all.

It will probably sustain a public sphere, and also leadership roles that may be formal or informal, which will shape public opinion within the community, and also represent it to outsiders. If the members care about the community, there will be debate about its direction and rules, and enduring factions may emerge. Community members may also have obligations or responsibilities to their communities; one typically owes loyalty to more than one community, and may have to make decisions about priority or compatibility between them.

The point is that members feel belonging to the community, and connected to each other. They needn't have a great deal in common, necessarily. The strength of their perceived connection may wax and wane, perhaps related to interactions (good or ill) with outsiders.

CONSERVATISM, COMMUNITY AND TRUST

Communities are home territory for the conservative thinker. Even the founding fathers of liberalism tacitly assumed that members of a political community would share ethical codes and common sense views to provide a stable background to liberty (Urbinati 2015, 175). Nowadays, more advanced and "progressive" thinkers prefer to unpick community ties and ridicule their advantages, often in the name of fighting injustice or inequality in the presumed interests of non-members.

If we need to establish the goodness of our community objectively before we can admit to our deepest selves that it is what we love, what animates us, and what we will fight to defend, then we are already lost. No community is in and of itself better than another, and all communities judge themselves by their own internal standards. Being a member of a community means, if not to be uncritical, at least to drop one's critical guard, and to take pleasure purely and simply in its symbols, triumphs, and history. A member of a community sees herself reflected in it, deliberately eschewing forensic self-awareness, whereas the self-conscious sophisticate takes a more critical and cosmopolitan view, trying to navigate it from an outsider's perspective (Reemtsma 2012, 13). But the latter strategy is too superficial to sustain deep relations.

Communities provide the matrix for the satisfying sense of *belonging*: that feeling of contentment and security accompanying reciprocal ties to an external group of people (Mahar et al. 2013). Trust is the glue that turns a cluster of people into a "we," transforms a social or political association into a destiny in which its members are invested. Reciprocity flourishes, because members continue to associate after their transactional business is concluded, and have interests in sustaining the community and maintaining their reputations within it. The community is a ready-made audience of beneficiaries for its members' trustworthy behaviour.

Trust is facilitated by community, thanks both to the legibility provided by shared understandings and reference points, and the possibility of sanctions in the event of betrayal (Lenard 2012, 85–86). It shapes the meanings of members' commitments – what is expected of a "good friend" or "business partner" or "police officer," for example, is not determined by the would-be trustee, but by the community in which she lives. It provides structures which can help people run or make sense of their lives, while also constraining and socialising them as well. It furnishes moral standards and social norms, and what Charles Taylor called *strong evaluations*, the "qualitative characterization of desires as higher or lower, noble and base and so on" (Taylor 1985, 23–24). Community supplies some of the most accurate social indicators of a trustee's trustworthiness, but simultaneously offers opportunities for duplicity. There is always a risk of being duped, even by "one of us," especially if the shared understandings lead to an excess of confidence which a trickster can exploit.

The basic quality of trust, the simple acceptance of the obligations of one social being to another, is fostered by community. Such trust is almost instinctive, quite possibly below the level of consciousness, through early socialisation or training (maybe even neurologically hard-wired). Many community activities are forms of life where fundamental mutual trust is a given. A community can, therefore, be undermined by untrustworthiness of, or lack of trust between, its members, and on some occasions it may undermine trust itself, for example by valorising the wrong kinds of relationship. Mere communality is not a panacaea for trust by any means (cf. Lenard 2012, 83–85). Not everywhere is rosy; some communities are characterised by crime and broken windows, and some by snobbery and complacency. Some are scarred by past betrayals or civil conflict. Many, pleasant and unpleasant, are allergic to outsiders.

THE DIVERSITY PROBLEM

It is a cliché of our time, cut-and-pasted into a million mission statements, that we "celebrate diversity." We do not, as a rule: people are happier in less diverse communities, and seek out homogeneity when they move, a pattern familiar in the concept of "white flight," but by no means restricted to white people (Crowder 2000; Woldoff 2011; Wilson 2019; Dinesen et al. 2020). "Homophily, the tendency to associate with similar others, is among the most robustly documented social phenomena" (Ertug et al. 2022, 38). Millennials are terribly embarrassed by the celebrated TV sitcom *Friends*, wildly popular in its day, because all its main characters are white (Cobb 2018), but social science testifies to its realism.

Diversity is often associated with its ethnocultural dimension, which has been seen as particularly problematic for democracy (Chua 2003; Mounk 2022, 1–84), and many have identified "the most salient ascriptive identities, like race and religion" as particularly divisive (e.g. Mounk 2022, 21). I don't want to single this dimension out; while it may be most associated with dangerous political flashpoints, the issues raised about trust apply more widely.

Diversity has become a target that all decent democrats believe they should aim for. Yet it is not, in itself, a value, and its value to a community will be highly nuanced and dependent on many aspects of its internal and external relations. It is undeniably valuable in many contexts, for example in knowledge-producing environments like universities (critical voices challenge groupthink), companies (out-of-the-box thinking leads to innovation), or cities (unfamiliar experiences breed new ideas, services and industries). Migration is thought likely to raise the incomes of both host and migrant (Goldin et al. 2011) by bringing fresh ideas. It is also true that diversity, where we have it, won't go away and if we don't allow for it and make it work for us, we will be worse off.

Many diverse communities are highly successful, and *entrepôts* such as London, Dubai or Singapore are characterised by a high level of diversity. Yet when the level of diversity changes, either through newcomers arriving or outgoers leaving, there may be shifts of power and social balance that undermine the *modi operandi* of the society, and affect the legibility of groups of people to each other. The Brexit referendum result in 2016 was not correlated with absolute levels of immigration, but rather the "leave" vote clustered in places with *recent* large-scale immigration (Clarke et al. 2017, 113–114; Goodwin & Milazzo 2017). The growth of sectarian political movements, whether of minorities or majorities (as with India's *Hindutva* ideology), can unbalance delicate compromises. Changing or democratising measures to protect the balance between diverse communities may themselves upset the balance, an argument commonly found in East Asia, where the racial mix can be volatile (Milne & Mauzy 1999, 80–102; Mauzy & Milne 2002, 99–113; Chua 2003; Tay 2017; Zuhari & Lidiqi 2019).

Sometimes, we have to accept that people do have diverse upbringings, reference points, experiences, values, languages, histories, political preferences, aesthetic standards, heroes, aims in life, educations, incomes, diets, and taste for Marmite, and live with it. Inter-community links may be tenuous, and mutual understanding relatively shallow. That doesn't make trust impossible by any means, but it makes it harder to avoid disappointment. In a diverse group, it is easy to trust the wrong person, or to fail to understand the other's intentions.

Not only is diversity a known problem for trust, a tight in-group with little diversity may be a recipe for success in some cases. Diversity of company boards, often a requirement of being listed on an exchange, doesn't seem to boost returns and may harm financial performance (Fried 2021). "Venture capital is dominated by white men drawn from a narrow combination of elite colleges" (Mallaby 2022, 384), but its positive effect on technology, the economy and society has been strong (Mallaby 2022, 375–379). The tech industry, stuffed with nerdy white males, often promotes diversity, but companies' growth periods, when they take the world by storm, are usually marked by its absence (Houser 2019; O'Hara & Hall 2021, 63–65). While a lack of diversity brings obvious problems such as the potential exclusion of talented people, or operational bias, sometimes a team of similar people with similar aims can more easily achieve a desired outcome. While diversity can reduce groupthink and bring in useful outside experience, this might be at the cost of coherence, legibility, loyalty and solidarity, affecting the willingness to put in twelve-hour days now for the possibility of a big payout down the line. Note that lack of diversity in technology is not a uniquely Silicon Valley problem; successful Indian and Chinese tech startups are markedly lacking in diversity across gender and ethnicity too (indeed, more so than the Americans).

Diversity is therefore an important issue for conservatism: how quickly is it changing, how much does it detract from mutual legibility and social cohesion, how much does it contribute to innovative thinking and prosperity? Conservatism takes diversity as a relevant social parameter, neither necessarily good (*contra* the Puritans), nor necessarily bad (*contra* the Peasants). A conservative would defend existing diversity in a successful and functioning community against attempts to reduce it.

As one scholar put it, it is an "open question … whether the increasing cultural and ethnic diversification of contemporary societies can lead to the formation of communities of belonging beyond communities of identity"

(Antonsich 2010, 644). Diversity has positive benefits, absolutely, but so has trust – it will depend on the community and its context. As often in the politics of trust, a circle has to be squared.

TRUST FACTORS IN COMMUNITIES

Wittgenstein wrote that "If language is to be a means of communication there must be agreement not only in definitions but also (queer as this may sound) in judgments" (1958, §242). In other words, communication is not simply a matter of agreeing on what words mean, but also on which sentences are true, false, absurd, appropriate, efficacious, too obvious to mention, even aesthetically pleasing. Competent speakers grasp these within an instant. A community is held together by its language games and the forms of its life that give it meaning. Those lacking understanding will find the community less legible, and will be less legible to it. That makes effective trusting harder and trust itself riskier.

It does not, of course, make trusting impossible. Communication is always possible, because no human is so remote as to be totally out of reach (Urbinati 2015, 27).

> Suppose you came as an explorer into an unknown country with a language quite strange to you. In what circumstances would you say that the people there gave orders, understood them, obeyed them, rebelled against them, and so on?
>
> The common behaviour of mankind is the system of reference by means of which we interpret an unknown language. (Wittgenstein 1958, §206)

This "system of reference" might exploit a number of commonalities in the human condition. Geography is important: however we live, we need to contend with the specific properties of space and place. Other near-universals are created by the body and biology – pain, the need for shelter, means of conceiving and protecting children. Some emotional demands, such as those for respect and privacy, are likely to be common. Maslow's hierarchy of needs, a (pseudo)psychological theory with little academic force, but an intuitive narrative of strangers' behaviour, is also suggestive (Maslow 1943; Abulof 2017). In short, although the notion of "human nature" is not fashionable, there is much we can glean when we "concentrate on normal people in a commonsensical way" (Kekes 1998, 35; Scruton 2017c). Hence Wittgenstein's explorer is not going to be at a total loss.

People don't need to subscribe to particular values, norms, belief systems or standards of behaviour to live together. A community is always evolving, and so such things cannot be fixed. People within a community may simply happen to live side by side, but they still need to communicate, and for that they need relevant things in common: "a collection of people who, in respect of their common recognition of a manner of attending to its arrangements, compose a single community" (Oakeshott 1991b, 56). The more they have in common, the better they communicate. We can talk in terms of a public culture which "garners respect in the eyes of members because it contains an important, present, active, and public dimension" (Lenard 2012, 77), or moral traditions which "unite their participants by providing them with a common way of interpreting and responding to the world" (Kekes 1998, 116). Such communication will tend to break down barriers between in-groups and out-groups, as bridges are built, prejudices crumble, and groups integrate in unpredictable ways (Mounk 2022, 231–252).

Some forms of life are valuable largely *because* they are shared (rather than their intrinsic properties), creating network effects whereby more people take part in them and find them legible, enabling a virtuous circle of richer communication across a wider population. Liberal criticisms that sharing and learning are unnecessarily exclusive and coercive on those who have to make efforts to join (e.g. Mason 2000, 117) merely threaten the value of a rich form of life, while denying minority populations access to a valuable source of social capital without any tangible compensation.

A community or society is exclusive to the extent that it demands that members subscribe to or respect common aspects deemed relevant and central. If it is inclusive, then members will generally extend trust to those who do not actively oppose or traduce common aspects. Ideologically, we can argue that one or the other approach is better. However, in practice, most societies maintain both approaches at the same time, swerving flexibly in the face of varying opportunities and threats, sometimes surprisingly quickly. One fact remains paramount: whether a society or community tends to trust outsiders or not, the practice of trust itself is aided by legibility and, all things being equal, trust will be placed more effectively and accurately when trustor and trustee are legible to each other.

In this section I will look at some of the factors that support legibility and are supported by communities' forms of life: shared values, common experience, collective memory and sources of authority. Trust is facilitated by all of these, though none is essential.

Shared Values

Shared values (moral or otherwise) are important within a community (Lenard 2012, 6), especially for its sustainability and survival. "If we create a society that our descendants will want to hold on to, our personal and political values will survive in them" (Appiah 2005, 137). Yet it is not all about "creation" and the future: being faithful to a community's *past* involves compassion for its tragedies, gratitude for its inheritance and sympathy for the values it held and nurtured (McCloskey 2006, 168–169).

Agreement on values is neither necessary nor sufficient for a community to flourish (Oakeshott 1975, 165–168). The history of artificial societies designed to promote particular moral, religious or environmental values is a comedy of short-lived, fractious exercises in fissiparous self-absorption (Trahair 1999), while utopian social writing quickly descends into jargon-strewn hallucinations of performative emancipation from hegemony, and ironic co-creation in vibrant, decolonialised hubs (e.g. Beauchesne & Santos 2017), without much consideration of who is going to make the tea or, indeed, get the electricity through the wires to power the kettle. The tendency of those with shared values to fight with each other over detail means that they might actually be counterproductive (the narcissism of small differences). Utopian communities struggle because their participants lack genuine community spirit; values can't replace it. After all, they are generally made up of people who had no use for the communities that nurtured them; they decided not to fit in there, or were unable to. Old habits die hard.

It is likely that value systems, such as global religions, are likely to present themselves differently in different societies, conditioned by history, geography, economics, religion, politics, geopolitics, and even national character. They may prove more problematic when they are recent imports, veering toward fundamentalism in the absence of venerated methods of social restraint (Roy 2004, 117–147). Even self-consciously global, universal values will eventually take on local forms. For example, progressive liberalism has developed differently even in societies as similar as Denmark and Sweden (Geugjes 2021, 227–228), becoming a tradition in each country rather than a universal principle, and not coincidentally mediating interactions and trust more effectively *within* the local community, familiar with its meanings, than with the *outsiders* it aspires to welcome.

Values that are shared may themselves be more or less conducive to trust. The more important values are relatively naïve, vague and common-sensical. Rough agreement on what is good, useful, beneficial, functional or positive, and bad, pointless, harmful, dysfunctional or negative in real-world contexts is often evident in unreflective discourse (Hollis 1998, 156). Once secure in our instincts, big picture arguments can rage about the various merits of justice, prosperity and equality (Letwin 1999). Ideology matters less than a community's intuitions about what is at stake.

In particular, people have a small-c conservative bias toward their existing habits and usual behaviour. This has obvious functional utility, because it renders society predictable. But given that virtue and values are often expressed through habit and routine, preserving behaviour will indirectly preserve values – we do not need to have an explicit attachment to them, nor to impose them on those whodo not share them. Trust and trustworthiness both depend, not on any sophisticated value, but a basic belief in the social value of trustworthiness and willingness to continue traditions of trusting and trustworthy behaviour.

One ethical concept that has been lost in the modern individualistic era is the importance of preserving the general order of things, as noted in Chapter 3. Ancient concepts, values and virtues such as *Ma'at* (the Ancient Egyptian goddess who personified the idea of immanent truth, balance, harmony and justice) used to express and convey a common-sensical view of how society ought to be. Shakespeare's tragedies often trace the baneful effects of transgression from social and hierarchical commonplaces (Lear abdicates his rightful position; Macbeth and Brutus are regicides; Othello doesn't trust his wife; Hamlet doubts his father). The idea limped on into modernity with concepts such as Burkean order and Hegelian *Sittlichkeit*, but these have also died out as common political currency (although see Bushnell 2022 for an attempt to revive order as a conservative aim).

Yet consensus about social order, I believe, persists, uncelebrated and now unnamed, helping to explain the attraction of policies that go against the most "advanced" social-scientific thinking, such as workfare (people do not deserve handouts if they haven't contributed to society), immigration controls (people whose forebears have built a society deserve more of a say in its future) and minimum wages (people should get a decent day's pay for a decent day's work). The Ancient Egyptian was surely wise to associate righteousness with respect for the continuity of the community, and *Ma'at* is contrasted with *Isfet*, usually translated as "chaos" or "strife." "'Ma'at' is the principle of social and temporal connectivity. It keeps time and society together. … Ma'at is not only a body of prescriptions and norms, but also a promise of duration and immortality. … A virtuous life bestows permanence to a person so that he/she may live on in this memory-space of permanence" (Assmann 2004, 10).

As well as deep ancient ideas, shared common-sensical values furnish benign behavioural signals. Uslaner made the acute point that an important attractor is a shared sense of humour and the ability to laugh at ourselves and our opponents: a "collective sense of humor that is so necessary to sustain cordiality and the trust in others that is essential to reciprocity" (Uslaner 1993, 1; cf. Wittgenstein 1998, 95; Neiman 2023, 11–12). Socialite MP Chips Channon expressed his failure to click with his boss Rab Butler by saying "Unfortunately perhaps we don't laugh at the same jokes" (Channon 2021, 836). One recent symptom of the recession of warmth, sympathy and humour is the outrage industry that has exploited X/Twitter in particular (Yarchi et al. 2021), for which Peasants and Puritans blame each other (Ham & Benson 2015; Kelly 2020). Outrage doesn't preclude humour, but it restricts it to ridiculing opponents and perking up supporters; different styles predominate across the divides (Berry & Sobieraj 2014; Young 2020).

The lesson of all this is that trust is prior to political, ethical or ideological value. Alice has a set of values, Bob has more or less the same set. All well and good. They are highly legible to each other, so Alice can make a fine-grained set of commitments that Bob can appreciate with discernment, and they can negotiate with precision. Yet Alice may still be untrustworthy, and/or Bob may still evaluate her trustworthiness negatively, or refuse to place trust in her, or make unrealistic demands of her. A shared set of values is useful, but hardly the whole story; no complex society will furnish a consensus (Crick 1964, 176).

Common Experience

The negotiation inherent between the trustor's needs and the trustee's commitments will also be eased by common experiences between them. Familiarity with a community brings with it understanding of the ways in which people interact, their customs, traditions, habits, etiquette and expectations, and their easy negotiation of landscape and geography. Values or ethnocultural identities are abstracted and remote from daily life; far more important are appreciation of landscape, architectural styles, churches, mosques or synagogues, the weather, meals, drinking habits, gender distinctions and other practical knowledge that people are familiar with and hold in common. These may be enjoyable in themselves, even if less enjoyable than outsider alternatives. Fish and chips from a newspaper doesn't really compare with the finest Italian cuisine, but it's jolly nice all the same and I wouldn't want to give it up. Britons have had to surrender the newspaper to the relentless forces of health and safety, but even the impure experience of greaseproof paper and polystyrene boxes (far more polluting) is worth it.

Common experience will typically be accompanied by a feeling of ease with one another, which might predispose those within the community to favour each other, and to respond to sympathetic social pressure. We "find it very difficult to follow [our] own reason or inclination, in opposition to that of [our] friends and daily companions. … A good natur'd man finds himself in an instant of the same humour with his company; and even the proudest and most surly take a tincture from their countrymen and acquaintance" (Hume 1978, 316–317).

The value of common experience is obvious: it enables legibility and communication. From it follow all sorts of equally valuable things, such as empathy, sympathy, imitation of others, adoption of norms, and the skill of experimenting within the bounds of a tradition. Radical thinkers often worry whether this will "generate unpleasant phenomena, such as an uncritical adherence to the most widely shared tastes or the burying of individual judgment in the opinion of the majority, or the imitative adherence to intolerant and discriminatory models of behavior" (Urbinati 2015, 26–27). It is not explained why these are unpleasant; they are not obviously so, and imitative adherence may also be to tolerant and inclusive models of behaviour. However, even that author celebrated "the individual naturally prone to respond to others, capable of associating, of feeling a sympathetic proximity with someone different, of identifying with someone in need, of cooperating … for the sheer pleasure of it; finally, of feeling an imaginative proximity … with all human beings, even those who are distant and far removed" (Urbinati 2015, 27).

Common experience does put outsiders at a disadvantage, at least initially; they have to pause, take stock, consider what regularities are detectable and infer or intuit rules that they can then try to obey, being open to correction at all times. And similarly, insiders will have to work to decide which rule the perhaps bumbling outsider is trying to follow, if it is not yet obvious. All this makes making commitments, signalling commitment and satisfying trustors' expectations trickier.

But not impossible. Our common experiences depend on practices or activities spread through our social lives, such as good manners, gossip, fashion, traditions, appreciation of art, love and marriage, clothing, sports, agricultural practices, food and ways of consuming it, our relations with animals and nature, hospitality, family connections, contracts or business arrangements, the law and the relationship with the state. These activities involve cooperation and agreement, sociability and empathy, copying and learning, and can be deeply embedded in small communities, but, as can be seen from the list here, they are also capable of being shared across heterogeneous groups. Where there is difference, commonalities we do have (for instance, living close together in a global city) can be reference points to help make sense of each other.

Commonality, in any case, is relative to a perspective. People in the same community have a lot in common, naturally. But what about rich and poor? Perhaps wealthy people in different communities have – in some respects at least – more in common with each other than they do with their poorer brethren, and *vice versa*. How about men and women? Those of different ethnicity? The able-bodied and those with disabilities? The beautiful and desirable, versus the plain and ordinary? The educated and the unsophisticated? The religious and the secular? The Bohemian and the traditionalist? The honest and dishonest? The young and the old? The famous and the unheralded? The gym bunnies and the couch potatoes? How much of a woman's experience can a man understand? Can someone whose body is his temple understand the profound pleasures of a night on the booze? Even those with plenty of experiences in common may find themselves baffled by their fellows.

The basic trust reflex, the acceptance of the obligations of a social being, is easier and supports so much more interaction when commonality and legibility are stronger, particularly where the commitments being made are complex and context-dependent. It follows that those lucky enough to have the benefits of commonality should eschew radical ideologies or practices that reduce or "hollow out" common practice (Gray 1995, 112–117). But common experience, like shared value, is not essential to trust. It may be that trust without commonality is shallow and broad, but hardly negligible for all that. All trust has to start somewhere, and a search for what we have in common is a decent beginning.

Collective Memory and History

The past gives rise to another important commonality, of joint antecedents. The preservation of the past in the present is a guide to "what works," defeasible but more reliable than the special theories of an individual or interest group. Cicero argued that Roman government was successful because it was "based on the genius, not of one man, but of many; it was founded, not in one generation, but in a long period of several centuries and in many ages of men. … [T]here has never lived a man possessed of so great a genius that nothing could escape him, nor could the combined powers of all men living at one time possibly make all the necessary provisions for the future without the aid of actual experience and the test of time" (Cicero 1928, 113), as Burke (1999a), Michael Oakeshott (1991a) and Hayek (1945) would agree. Beyond the practicality of relying on existing structures, there are sentimental considerations of continuity and identity too, the wish "to derive all we possess as *an inheritance from our forefathers* … carefully formed upon analogical precedent, authority, and example" (Burke 1999a, 119, his emphasis). Yet this sentimentality has its practical side too: "the idea of inheritance furnishes a sure principle of conservation, and a sure principle of transmission; without at all excluding a principle of improvement" (Burke 1999a, 121–122).

Collected wisdom, expressed in memory, history and practice, depends crucially on human psychology, but also on a social context of language, emotion, communication, action and interaction, to give it shape and structure (for the *locus classicus*, Halbwachs 1992). Recall may be a matter of habit (Turner 1994), or transmission through education and the media (Anderson 1991), or procedures, rulebooks, traditions, ceremonies, rituals and even public buildings. Such spatial, temporal and environmental mnemonics keep the collective memory, which may easily be lost if not directly relevant to daily life, in community members' consciousness (Assmann 2006a).

Socialisation is essential for memory, but the converse is also true. "Socialisation is not just a foundation, but also a function of memory" (Assmann 2006a, 4). Once our social memories are established, we take our part in the community more readily. We perceive ourselves as members; remember our status, see that we belong. We share a semantics of the past that is often (not always) unifying. In Jan Assmann's judicious phrase, collective memory is also *connective* memory (2006a, 11).

The social aspect of memory is vital for trust, most obviously because trust depends crucially on memory. Where Alice has made a commitment to Bob, they both need to remember it, and accurately. Friedrich Nietzsche blamed memory for the imposition of conformity, banality and (worse) responsibility for precisely this reason. Whereas "forgetting is a strength" for the Nietzschean *Übermensch*, he berated "memory, with the help of which forgetfulness can be suspended in certain cases, – namely, in those cases where a promise is to be made" (Nietzsche 2007, 36). While Nietzsche, like many of the Puritans and Peasants whom he has (banefully) influenced, could see only how responsibilities restrict our freedom, more sensible thinkers note the advantages that "ontological security, the trust that everyday actions have some degree of predictability" (Giddens 1984, 87) provides, underpinning both individual identity *and* membership of the community.

Continuity through time gives meaning and form – some activities take decades, and are not easily deducible from the actions that implement them. Consider getting a pension. In one sense it's a matter of signing a form and making payments, but the whole activity involves working, making sacrifices in the present, maintaining a level of income which sustains the payments, and so on. Each one of these aspects of "getting a pension" has its own trust requirements. It also reaches back to the business of pension management, itself based on trustworthy payment of annuities, which has grown up as one of the many ways that capitalism enables value to be stored and accessed at a later date.

Wittgenstein gave the example of a country created instantaneously that is a perfect copy of a part of England, that survives for exactly two minutes. Everyone in the two-minute world is "doing" the same thing that their English counterpart is doing in the subsisting world at the same time, until they vanish into the ether again. Do we really want to say that a two-minute person is working for his pension as he sits at his desk tapping away at his computer? "Could we for example not imagine a past and a continuation of these two minutes, which would make us call the processes something quite different?" (Wittgenstein 1978, VII §34). Or even: how can we call it "working for his pension" when he has no bank account, no history of capitalism, no money economy, no tradition of respect for the elderly, no ambitions to shake all cares and business from his age, conferring them on younger strengths, while he unburdened crawls towards death?

Trust across generations is important, as Burke famously argued. A state, a nation, is a partnership, and "as the ends of such a partnership cannot be obtained in many generations, it becomes a partnership not only between those who are living, but between those who are living, those who are dead, and those who are to be born" (1999a, 193), building on but not restricted to "the relation of blood [which] produces the strongest tie the mind is capable of" (Hume 1978, 352). We hold the world *in trust* for future generations, and should have sympathy and gratitude to our ancestors for tending the world we have inherited. "Under a pious predilection for [your] ancestors, your imaginations would have realized in them a standard of virtue and wisdom, beyond the vulgar practice of the hour …. Respecting your forefathers, you would have been taught to respect yourselves" (Burke 1999a, 124). This is an ancient insight; *Ma'at* emphasises gratitude as a reaction to the past (Assmann 2006b, 90).

Walter Benjamin had a point that "There is no document of civilization which is not at the same time a document of barbarism" (2019, 200), but beggar-thy-neighbour politics badges historical events as good or bad, innocent or guilty, and the sense of gratitude is trampled in the consequent rush to judgment. In his study of the United States, Uslaner discussed the 20th century experiences (2002, 37, 161–162), which affected generalised trust positively (such as the civil rights movement) or negatively (such as Vietnam). But collective memory is always in dialectic with the present, and the civil rights movement is far more polarising now than when Uslaner was writing; some see it as an illicit power grab by activist Supreme Courts, whereas for others it is unfinished business, as evinced by the murder in 2020 of George Floyd. Uslaner also argued that the greater equality and optimism of the baby boomers increased social trust (2002, 162, 181). However, nowadays the boomers (of whom I am just about one) are blamed for a number of ills, including borrowing from future generations to increase their own living standards, and presiding over environmental catastrophe (Willetts 2010; Gibney 2017). So much for holding the world in trust. From today's perspective, it looks like the boomers, when in the majority, were able to create the illusion of generalised trust, but actually it was the particularised trust of an in-group. Now that different generations have emerged, throwing abuse at each other, Uslaner's picture looks less rosy.

History and memory, like shared experiences and values, can underpin legibility, but interpretations of events can change. Far from showing sympathy for or gratitude to our ancestors, their actions are now brought into the realm of present public opinion and judged anachronistically by modern standards (often themselves contested). Meanwhile, communities are increasingly taking in diverse outsiders whose relation to that history may well be problematic. In the absence of sympathy, history may become a divider, and itself a cause of mistrust (Assmann 2006a, 21–24). We may need to adjust our histories and memories to become more accepting, or to dispel mistrust, but equally this cleansing cannot be done without sympathy. History and memory are essential for the sustainability of the community, and we cannot simply tear them down and re-erect another edifice, least of all one in which our ancestors are condemned without mercy.

Authority and Hierarchy

Our experience of a community depends on our position within it. With this in mind, we should consider arguments from Uslaner and Seligman that hierarchy and inequality are inimical to generalised trust. I have already argued that generalised trust is neither realistic nor helpful, but if their arguments extend to more practical forms, then conservative politics will be in trouble. Conservatism supports both particularised trust and hierarchy; if hierarchy in turn rules out particularised trust, then we won't get anywhere very fast, and we will be condemned to the atomised, individualistic cosmopolitanism that is the end point of Uslaner's project.

> If we believe that we are connected to people who are different from ourselves and have a moral responsibility for their fate, we see that trust is a fundamentally egalitarian ideal. When we take others' moral claims seriously, we are treating them as our equals. A belief in hierarchy is inimical to moralistic trust. A culture of trust depends upon the idea that things will get better for those who have less and that it is in our power to make the world better. (Uslaner 2002, 3)

This is based on two arguments, that "optimism for the future makes less sense when there is more economic inequality," and "if there is a strong skew in wealth, people at each end may feel that they have little in common with others" (Uslaner 2002, 38 – I have already acknowledged the possibility, but not the necessity, of the second claim). It follows from these that only generalised trustors "give of their own time and money to help the less fortunate [and] support governmental policies to redress social and economic inequalities" (2002, 192). A "rise in inequality may well tap an increase in materialism among those who are well-off and a corresponding decline in the public-spiritedness that undergirds trust" (2002, 207), because "if you believe that economic stratification is justifiable, then you have no need to trust those below you on the economic ladder" (2002, 86). Across the globe, "the most important determinant of trust is the level of economic inequality" (2002, 230).

This nest of arguments is rather odd. In the first place, trust and hierarchy almost always go together. The student and teacher, the violinist and orchestral conductor, the athlete and coach, all sit in voluntary hierarchical association to mutual benefit. In each pair, the latter has authority, because of her experience, but has no power to compel the former to obey her (Kekes 1998, 137–139). Of course, these are not the types of trust that Uslaner was interested in – they are self-interested and "strategic." Yet they are the very bedrock of society and the little platoons of which it is made up. Any nation built on institutions like the family and freedom of association will experience inequality, because some families and some associations help people, and others hinder them (Timothy 2020, 191). There isn't very much we can do about the importance of family, and the inequalities it introduces, apart from going down Plato's *Republic*'s route of removing freedoms and putting everyone under a central and standardised authority – trustless trust under a philosopher-king. On the other hand, many enjoy the extension of familial hierarchy to the state itself – in Britain, the Netherlands and other monarchies, royalty is popular and only really disliked by intellectuals who think they should be in charge.

Conversely, the moralised and non-hierarchical relationships between equal participants that Uslaner championed are important but hardly core to community. "Relations between equals and nonintimates will *be* the moral norm for adult males whose dealings with others are mainly business or restrained social dealings with similarly placed males. But for lovers, husbands, fathers, the ill, the very young, and the elderly, other relationships, with their moral potential and perils, will loom larger" (Baier 1994e, 114, her emphasis). Indeed, "equality is not even a desirable ideal in all relationships – children not only are not but should not be equal in power to adults – and we need a morality to guide us in our dealings with those who either cannot or should not achieve equality of power (animals, the ill, the dying, children while still young) with those with whom they have unavoidable and often intimate relationships" (Baier 1994e, 116). Attempts to eliminate hierarchy and impose equality have the perverse effect of engaging the state to unpick these fundamental relationships (Timothy 2020, 176–177).

Moving from survey to theory doesn't improve the argument. Seligman (1997) argued that most of what we call "trust" is really confidence in a rigid social system, and trust only emerges in modernity as belief in the goodwill of the other as she exercises her agency within different roles. It followed, for Seligman, that trust as such cannot take root in a hierarchical culture with rigid social structures. Yet what he claimed distinguishes modernity from the premodern is the proliferation of roles; we occupy more of them than our premodern forebears did. Indeed, but roles are typically hierarchically designed – almost any role you can think of, especially in the public sphere, delineates a set of powers (downward) and obligations (upward) relative to other roles. Because we occupy more roles in the modern world, we take our places in more hierarchies. There is no Great Chain of Being, nor a rigid feudal hierarchy from king down to peasant, but there are many organisational hierarchies, often quite fluid, and if they are destructive of trust then we have a problem.

Of course, hierarchy doesn't destroy trust, although contexts where power is wielded hold temptations for the unworthy. Ithiel de Sola Pool's suggestion that a society that rejects status as a source of credibility, particularly cognitive credibility, will have an unusually great need for trust (de Sola Pool 1963;

Luhmann 2017, 65n.14) has proven to be prescient as liberal dismantling of hierarchy and authority has led to a misinformed, conspiracy-theorising world.

Trust in authority is vital in all sorts of areas, including the transmission of knowledge and the operation of expertise. Such trust must rest on the trustworthiness of those who occupy authoritative positions. This is at least partly what differentiates hierarchical authority from the mere exercise of power (Luhmann 2017, 57–58). Yet this is routinely denied in modern political thinking, which uncritically assumes that the ideal of the dignity of the individual brings with it "contestation and resistance to the constituted power, irrespective of whether it is legitimized by the opinion of a majority that has been elected or imposed by the arbitrary will of a monarch" (Urbinati 2015, 14). Neither majority opinion nor a monarch's will is necessarily bad, which affords the dignified individual another option: acceptance of properly constituted power.

COMMUNICATING COMMON INHERITANCE TO OUTSIDERS

Shared values, common experiences, customs, traditions, memories, histories and hierarchies are supportive of, though not necessary for, trust. When Alice and Bob have much in common, it shapes their outlook, including what he demands of her, and what she is prepared to commit to. All things being equal, their intentions will be legible to each other *in their context*, giving them reason to work together to sustain the context. Bob won't ask too much of Alice, and she won't provide too little. As their contexts are partly set by the *communities* in which they are embedded, their joint interest means that they each have incentives to sustain those as well.

If a certain practice is important within a community and depends on both trustworthy behaviour and reliable recognition of trustworthiness, then Bob and Alice will have a learned appreciation for what will sustain it. Their signs, signals and language may be impenetrable to outsiders, but they are versed in the practice and prepared to make efforts to foster it. For an outsider to benefit from the value that the practice affords the community, he or she will have to take steps to learn and adopt it, or at least adapt to it, and not undermine it. On the other hand, all things being equal, any community will benefit if its members positively seek trusting relations with outsiders, and engage with those unfamiliar with its customs with sensitivity and sympathy.

The Cognitive Component: Citizenship Tests

This can be easier said than done. Consider the following flawed argument:

> As longtime members, we may experience participation in the public culture as natural in some sense, and therefore we may think of the trust we extend to comembers as in some sense intuitive or unconscious. Yet that it is in fact *information* that can in principle and in practice be imparted to newcomers tells us that the norms and values that define a public culture have an important cognitive element, and can be learned and adopted over time by those who join us. Long term members may focus on what "we" have in common, and therefore attempt to harness the emotional or sentimental foundations of trust, but in integrating newcomers, it will be essential to focus on the cognitive underpinnings of trust. (Lenard 2012, 90, her emphasis)

The phenomenology of participation and belonging cannot be so easily set aside. The representation of the values, experiences, memory and hierarchies of a community in the form of information is neither complete nor sufficient, though it is of course a start. The assumptions in this quote underlie the drive to develop "citizenship tests," often multiple choice, where foreigners demonstrate their understanding of the host. Cognitive underpinnings are the natural place to start in citizenship education.

Yet even if the essence of a community can theoretically be distilled into information, it is not obviously practicable. Firstly, the cognitive effort required to internalise a community's norms, or alternatively to retrieve them from a learned list, may be too great to be of use in a real-world situation (as opposed to an exam). Secondly, newcomers come from communities of their own, and retain their own norms, which, as they are also information by hypothesis, may clash or be inconsistent with the new set, interfering with recall. Thirdly, there is no reason to think that the newcomers have a complete command of the concepts with which our norms are expressed. Fourthly, there is no reason to think that the community itself can set out its own norms with confidence, that there will be wide agreement on how to express them, or even wide agreement on what they are (consider the Peasants' and the Puritans' very different views of what Western democracies are about). And fifthly, the evidence of actual citizenship tests tends to show only the randomness of their content (Michalowski 2011), putting in doubt the claim that a definitive curriculum can ever be developed.

Given goodwill of both insider and outsider, there is no reason why trust should not burgeon. A connected community with valued ties and solidarity can provide the foundation both to accept outsiders and to maintain its identity as it does so. Clearly not everyone has to hold the same set of values, or subscribe to the same cultural memory list; people can and do challenge authority and hierarchy from within the community all the time (indeed, the nostrums of a valued, living community are likely to be robustly contested). Furthermore, public values are likely to be fluid, evolving with social change, so, while they are certainly a centripetal force, they are not homogenising, and neither will they eliminate diversity (Lenard 2012, 76). They are likely to adapt to incomers' mores in time (Lenard 2012, 82), although it is desirable that this should happen slowly to retain legibility.

It is also important that such change should happen from the bottom up, pulled by the forces of social evolution, and not imposed from the top down, pushed by rationalist Brahmin theories of how society "ought" to be. It is reasonable to say that the public sphere should be free of discrimination and prejudice; however, it is not possible to legislate that the private sphere should be similarly free. Interference from the top down does not regulate the private sphere – it only shrinks it (O'Hara 2023, 267–272).

The Interests of Incomers

Western academics are quick to blame majority communities for trust failures. For instance, "The discrimination faced by cultural minorities is often a consequence of a lack of trust in one direction, that is, from majority to minority; the consequence of this discrimination is equally a lack of trust in the other direction, that is, from minority to majority" (Lenard 2012, 3). Mistrust from the minority is framed as a rational response to its own mistrusted status (Mounk 2022, 104).

This cannot be right, if for no other reason than a minority, being both smaller and an arrival in a society that it has not built and in which it has less stake, needs to be accommodating if a *modus vivendi* is to be reached. Mounk made the startling statement that cultural change driven by an influx of minority groups "is likely to … cost groups that have long stood at the top of their countries' ethnic and religious hierarchy some of their unearned advantages" (2022, 172). Unearned? This is extraordinary. In the first place, if the country is successful enough to attract newcomers, then those on top of the hierarchy might well argue that they deserve any advantages they have. They have built an appealing society, and understand it well enough to thrive in it. Why should this not have been earned? Conversely, if Mounk is correct to say that building a society doesn't earn you any privileges at all, then it surely follows that merely entering a society as a minority outsider doesn't earn you any either. Wherever the minorities stand in the hierarchy is unearned, on this principle. If the indigenous population deserves nothing, neither does anyone else.

It seems far better to say that, if the wider society has proved attractive to the minority that wishes to join it, then it is not in the minority's interests to change it very much. Hence, complaints about seeing rights within the society "as privileges of those who already enjoy them and in the specific forms in which they enjoy them, or as rights of a cultural and political majority that happens to live on a given territory" (Urbinati 2015, 16) are misguided. The rights of the minority might well be better protected by a regime that attempts to shore up an attractive culture. A deliberately neutral regime may fail to provide a steer as to what counts as good behaviour around here. Furthermore, few groups "happen" to live anywhere; the successful majority has staked a claim to its brand of social organisation in the vicinity over generations, while the minority of arrivals were presumably attracted by something about that brand. At a minimum, it is presumably better (safer, or more prosperous, or more tolerant) than whence they came.

Puritan analyses of what they see as the regrettable tendency to prioritise one's group(s) over others have generated the academic construct of *collective narcissism*, a deliberately pejorative name for extending the self-regard of individual narcissism to one's own in-group (Golec de Zavala et al. 2009; Golec de Zavala & Lantos 2020). Certain instances of collective narcissism are clearly wrong, such as the promotion of one ethnocultural minority over others (as, for example, in *apartheid* South Africa or modern day Kenya), or the imposition of supposedly universal values onto indifferent or unwilling communities, for example the values of the global Muslim *ummah* or Western liberalism onto the global South.

However, in the specific case imagined here, where a group has built and maintained an attractive community for a certain period of time, and is losing its comfortable numeric dominance thanks to an influx of those who wish to share the community's benefits, communal self-regard could be compelling to the majority and would have an independent argument behind it, grounded in the appeal of the community to the incomers. Such a situation could foster collective narcissism and, who knows, maybe the narcissists would have a point.

Indigenousness

These considerations carry less weight if the minority has been present for a long time. If its interests are ignored, then there will indeed be risks of social cleavages (Mounk 2022, 104), and in such a case the minority has its own rights of territory (it too "happens" to live there). There will be complexities of history to unpick, historic injustices, economic co-dependencies, and so on. The majority will be foolish to ignore or disparage the interests of the minorities, on practical, political, economic and moral grounds, but its position as majority is not an insignificant one. Much depends on the quality of life, and the potential of any alternative – stamping out injustice is important as long as something better takes its place. It also depends on the ability of both sides to engage in debate and discussion, to present their alternative experiences and perspectives persuasively (Warren 1999, 340).

Where the minority is the indigenous community, as in Australia, Brazil, Canada, New Zealand, the United States and elsewhere, the argument of the previous subsection doesn't apply at all. I have much sympathy for such communities, and the balance to be struck is harder. The pragmatic argument will rest on the quality of life across the nation as a whole, as well as within the indigenous minority. Disparities between the two need to be addressed, but there is little consensus on the best way of doing that. It will always be important not to undermine the quality of life in successful democracies such as those listed, but it is unlikely that pressure from indigenous minorities will threaten that. These political problems can only be solved by getting-things-done politics based on trust between the communities, which requires hard work and resisting the temptation to sloganise on both sides.

Achieving Trust

The process of achieving trust between majority and minority (whether indigenous or immigrant) has to be two-way, and bottom up – but then it has to be accepted that meaningful interaction may not happen at all. Unforced interactions can't be *made* to happen, only *encouraged*. Only common projects, however small, and the cooperation and debate they engender, will achieve them. That will require the organic development of new in-groups with some genuine interest to hold them together, which will not be accomplished by a theoretical, unconditional and indiscriminate generalised trust of everyone – since if we knew how to create that, we wouldn't have the problem in the first place.

Common experience can't be created simply by waiting for people to mix, but neither can it be produced by setting up artificial forums or "collaborative spaces," however "vibrant." The metaphor of "astroturfing," creating an artificial playing field devoid of grass roots, is appropriate for these. The production of commonality is an organic process that unfortunately sometimes doesn't get properly started. Neither the Peasants' favourite idea of forcing minorities to assimilate to the majority, nor the Puritans' of remorselessly attacking the majority, is a likely means of increasing interaction or trust – resentment perhaps, but not trust.

Conservatism is the only ideology whose *raison d'être* is to ensure that communities, institutions and practices that have persisted and created social value in the real world survive the untested critiques and impractical schemes of those who would prefer something more modern, or "moral," or rational, or efficient, or utopian, or who would destroy such institutions in the name of "inclusion."

KILLING COMMUNITY WITH IDEALISTIC THINKING

Trust is important for a community: community fosters trust, and members are usually more willing both to extend and to carry out commitments to other members. There is, therefore, a potential danger that the drive for inclusion via generalised trust and cosmopolitanism will harm rather than benefit the communities it is intended to improve.

Trust and Democracy

To begin with, we need to dispose of an argument that trust must be generalised if it is to underpin democracy. In other words, trust in a democratic community must routinely and automatically be extended to all. That means that, to the extent that a community fosters particularised trust and loyalty to little platoons and other in-groups, it cannot be democratic. It is important to debunk it, if only because all communities contain such loyalties, which are valuable in their own right.

As articulated in recent times, the argument has two components. "First, citizens must believe that the unfair restriction of the rights of some individuals is an affront to the community as a whole" (Lenard 2012, 71–72). Second, political actors and citizens "must trust that the citizens they are aiding would be willing, if the need arose, to act on their behalf in return" (Lenard 2012, 72). Neither of these claims is ultimately convincing – what analysts call a "full democracy"[1] could easily function in a community with significant and competing loyalties.

The first component is a sleight of hand: citizens will indeed disapprove of unfair restrictions – everyone dislikes unfairness. The problem is the prior question of what is unfair. For example, indigenous members of a community may believe that their rights are more important than those of newcomers or outsiders. Puritans think this is unfair, as everyone should have the same rights, independent of their contribution to the construction and maintenance of the community. Peasants think it entirely fair that natives are prioritised over interlopers. Both camps are opposed to what they believe is unfairness; they just disagree as to what is unfair. Incidentally, both are also guilty of inconsistency in the application of their principles: the Peasants are supportive of indigenous peoples in Europe, and utterly uninterested in those of the Americas and Australia, while the Puritans' inclination is the exact inconsistent opposite.

---

1 Defined by the Economist Intelligence Unit's annual democracy index as "Countries in which not only basic political freedoms and civil liberties are respected, but which also tend to be underpinned by a political culture conducive to the flourishing of democracy. The functioning of government is satisfactory. Media are independent and diverse. There is an effective system of checks and balances. The judiciary is independent and judicial decisions are enforced. There are only limited problems in the functioning of democracies" (https://pages.eiu.com/rs/753-RIQ-438/images/Democracy-Index-2023-Final-report.pdf, 66).

As for the second component, it seems unlikely that many social groups rely on others to back them up in a lively democracy. Indeed, part of the function of democracy, and part of its attraction, is that groups can form to define and defend their own interests. In a narrow sense, they should be able to rely on the minimal support of all to put their own point of view forward effectively – this is part of the basic commitment to democracy that most in the West make. In actual politics, they may get some support from other groups, sometimes because they make a robust and coherent enough defence of their rights to impress others, sometimes because they join effective coalitions, and sometimes because they make such a fuss that they can extract concessions. Depending which route is followed (it could be all three simultaneously), the amount of inter-group trust generated will vary, but it could remain very transactional.

The problem is that if individual rights are seen as precluding ties and loyalties to groups and associations, minorities might easily be overlooked or disregarded, because in a world of individuals the majority finds fewer obstacles to majoritarianism. This is the "soft despotism" of which de Tocqueville warned: "as each class closes up to the others and merges with them, its members become indifferent to each other and treat each other as strangers. Aristocracy had created a long chain of citizens from the peasant to the king; democracy breaks down this chain and separates all the links" (de Tocqueville 2003, 589). If so, democracy might function better, at least as a protector of minorities, where trust is more discriminating, and less generalised.

The Dangers of Generalised Trust

Generalised trust demands equal treatment of everyone, understood as individuals whose characteristics and character are irrelevant to the assessment of their trustworthiness. The conservative tradition, on the other hand, values certain existing social structures, hierarchies and institutions, which will naturally influence who gets trusted and why. I argued earlier that generalised trust was a disturbing ideal, and in this subsection I shall highlight the dangers it poses to community (see Chapter 4 for the definitions of the various types of trust introduced by Uslaner and others).

Trust of our fellows remains vital for community and is at risk from shocks to it. We must not conclude from this truism that if we trust indiscriminately, the "right" type of non-discriminating liberal community of individuals will form. There is a slippery slope from the view that reliance on group connections, strong ties and local trust is insufficient to build successful and prosperous societies and nations, which is roughly Fukuyama's position (1995), to the view that strategic or particularised trust is a negative handicap, which is Uslaner's (2002).

I should say right away that, as we saw in Chapter 4, it is both possible and common to place trust in people about whom one has an ambivalent attitude or relatively low trust. "Altruistic trust,' where the trustor places more trust in the trustee than is warranted by the available evidence (Mansbridge 1999), is a possibility, which will oil the wheels of the relationship between insiders and outsiders. It often makes sense to trust outsiders altruistically, especially if you are keen to establish a relationship, expand your trusted circle, or even for purely transactional purposes. The question is not whether it is a good thing (because it is), but rather what the mix should be with trust more strategically placed, and whether there is a duty to be altruistic in this sense (put another way, whether an insider is in the wrong, behaving badly, if he does not trust altruistically).

Altruistic, generalised and moralistic trust are associated with cosmopolitanism – and there is nothing wrong with that *per se.* A prosperous nation needs both tight-knit communities and cosmopolitan cities to grow, innovate and sustain relations with external societies. But when cosmopolitanism becomes the ideal, it not only threatens to crush the little platoons on which the whole edifice rests, but also removes the protections of individuals that small associations sustain, based on deep familiarity, coincidence of interest and even chippiness about outsiders.

Uslaner argued that merely associating with others won't make (generalised) trust happen. Joining groups voluntarily usually means engaging with similar people, which doesn't help with trusting unfamiliar strangers (2002, 2–3). He also claimed that membership of in-groups was likely to suppress generalised trust in strangers, and that those most satisfied with their in-groups were the least trusting of newcomers (2002, 105–106). From this, he concluded that generalised trust is a precondition of civil society, not a result of it (2002, 4–5, 198). This is a mistake. There is no reason why joining groups voluntarily should not increase *strategic* trust, even if it has no effect on generalised trust, so that trust, albeit of the strategic variety, *can* grow alongside civil society in a virtuous circle (Putnam 2000). It would be surprising if that had no effect on generalised trust, but Uslaner's analyses suggested it wouldn't. I am sceptical of these analyses, but Uslaner's cosmopolitanism led him to conclude that generalised trust should be increased, *even at the cost of the loss of strategic or particularised trust.* That seems highly counterproductive.

Those who valorise unconditional generalised trust approve of cosmopolitan societies at the expense of the particular: "We must have positive views of strangers, of people who are different from ourselves, and *presume that they are trustworthy*" (Uslaner 2002, 2, his emphasis). Maybe, but the more immersed in a community we are, the more difficult it is to absorb strangers, who after all don't share the deep connections we and our forebears have painstakingly built up. Shared values, experiences and memories need not be generated deliberately by a particular practice or institution, but might simply be absorbed as a side-effect of living, worshipping, playing or being educated together. In such a community, strangers have difficulty being trustworthy because they can't fully grasp the expectations of the community (this is an epistemological, not a moral, failing). If we have deep knowledge about a sub-group of people, it would be silly not to use it in our trust judgments, which must be correspondingly more accurate and precise than our judgments about strangers.

Uslaner took that view as marking a lack of civilisation. "When you only trust your own kind, your moral community is rather restricted. And you are likely to extend trust only to people you think you know. ... [You] are not agnostic about strangers. Particularized trusters assume that people unlike themselves are *not* part of their moral community, and thus may have values that are hostile to their own" (2002, 27, his emphasis). Well, yes, they may so assume, and they may indeed be right.

But it is not correct to say that your moral community is necessarily restricted, for two reasons. In the first place, any restriction in breadth may be more than compensated for by depth. The fewer the shared values of two people, the smaller the connection between them, and the less that each understands the other, and hence the less that each can do for the other (at least without a lot of discussion, explanation and cognitive overhead). On the other hand, two people with many shared values can depend on each other's sympathy, habits and instincts. The interactions thus facilitated are far less superficial. Uslaner was only interested in the *number* of people between whom shared values facilitate trust, but this reductionism ignores restrictions of the moral community in terms of what one might expect from it, and the richness of the interactions within it. The people in the broad but shallow community that Uslaner advocated may find that risk aversion and lack of trust is still the best strategy, because its shallow interactions and weak ties don't support a more adventurous attitude.

Secondly, it is also reductive to assume that one's moral community is a unity. Most of us play parts in several overlapping communities, each with central and peripheral figures, so that across our social networks we will share an arbitrary set of values, which will allow more or less trusting relations of different types, for different things, with different people. One might trust one's priest for some things, parents for others, best friends for others, teachers, the police, strangers on the street, shopkeepers, doctors, lawyers, teammates, politicians. A stranger may have still have a role to play in trusting encounters across a community – just not in every encounter or context.

At worst, making the "fundamental ethical assumption: that other people share your fundamental values" (Uslaner 2002, 2) could morph into a coercive attempt to impose your own values on others by placing trust in them that they are unwilling or unprepared to accept, and demanding certain types of behaviour as a *quid pro quo* for being accepted so magnanimously. It is unjust to place outsiders in a position, for purely ideological reasons, where they need to behave in a trustworthy manner, before they have had a chance to internalise the standards that count in the community via a gradual process of learning, training and assimilation. Rather than remain content in such an awkward spot, outsiders may prefer to celebrate their own difference, displaying types of behaviour that you don't understand, or are unlikely to accept. They may feel the need, for their own self-respect, to push back when their own values feel eroded.

Social Capital Theory: An Aside

Some of the literature discusses trust in terms of *social capital*, leveraging the metaphor of capital as a type of wealth that persists, and can be stored, accumulated and later invested for one's personal advantage. Robert Putnam described social capital as a feature of social organisation which explicitly includes interpersonal trust (1993, 2000; Cozzolino 2011), while Pierre Bourdieu defined it as the set of resources linked to one's durable network (Rostila 2011; Bourdieu 2018). This view sees trust as a modern phenomenon (with Seligman 1997), characteristic of a rational and instrumental outlook, just as economic capital is a resource of the modern economy. Trust and trustworthiness are "delivered" in parallel more or less magically, if not without effort, by the functioning of the individualist liberal order, ameliorating its immanent complexity (Luhmann 2017, 27–35).

On the other hand, if trust is a more or less constant component of all functional civilisations, as I have argued, then it can hardly be a creation of liberal modernity, but rather something upon which it rests. The marvellous social capital of the modern world – which really is marvellous, after the tremendous leaps that Western society has made in the last three centuries – has not created trust and trustworthiness out of nothing, but rather presupposed the forms of global trust and trustworthiness that emerged slowly in premodernity, expressed in many ways, including proto-capitalist practices and institutions such as contracts and banks, complementing non-transactional types of local trust within groups, tribes, family and churches. Premodern, local trust didn't become obsolete (*pace* Reemtsma 2012, 13), but it wasn't scalable and failed to support wider religious, commercial and administrative projects.

The miraculous trust of strangers in the modern, globalised world, through this lens, looks more like the trust of relatively similar people, sharing or adapting to a long liberal capitalist tradition which grew out of earlier experiments, who perform trustworthiness in more or less similar ways. It does not follow that, if we treat people as atomic individuals of arbitrary cultural backgrounds, or hidden behind a Rawlsian veil of ignorance, trust will be as widespread, or trustworthiness as readily available. Let's take the metaphor of capital seriously: John Rawls' proposal of "nullify[ing] the effects of specific contingencies which put men at odds and tempt them to exploit social and natural circumstances to their own advantage" (Rawls 1999, 118; Urbinati 2015, 39) is effectively to destroy social capital, which is precisely the ability to arrange social and natural circumstances in one's own favour. In particular, it would remove the bases of trustworthiness signalling, and produce a society with less trust rather than more. The trust that remained might be generalised trust; so much the worse for that.

I don't object to the discourse of social capital, but I won't be using it myself; it is too neat an idea to hook up trust, trustworthiness and the accumulation of psychosocial "wealth" in a win-win for individuals and society. Social capital as expressed by both Putnam and Bourdieu is transactional, to the material benefit of the individual. This may indeed explain why rational people trust when there seems to be no reason to, and also why trust is by and large such a profitable risk, but they miss the point. Hume was far closer to the mark in his discussions of good nature and sympathy, and the tincture of company. There is profit to be made with good nature, but profit is hardly its only, or even its main, objective.

CONCLUSION: THE RESPONSIBILITIES OF INCOMERS

Someone transplanted into a complex web of relationships from outside may have problems getting herself trusted, especially if she appears different in looks, accent, language, or whatever, and possibly also if she has divided loyalties to other communities or a diaspora. In many cases this will be unfair and discriminatory. But the flip side is that our stranger may not understand the commitments standardly made within a society, and the non-verbal expectations that typically accompany them. The outsider may fail to be trustworthy, not because she is untrustworthy, but merely because she is unable to carry out her commitments satisfactorily (however willing she may be). She may simply have failed to register on anyone's radar. Trustworthiness needs to be not only manifested, but signalled, so the signalling system in play needs to be mastered.

It is important to realise that the problems faced by outsiders are not a contingent flaw in Western democracies that could be eliminated by a more welcoming or less morally-compromised society, as argued in this passage, for example:

> In practice public cultures often emerge from imperfectly inclusive processes; this is because the historical elements of a public culture developed in an era in which

> democratic inclusion was not a widespread commitment, and in part because democratic communities are not yet as inclusive as they ought to be, as a result of which minority members continue to struggle to interact on equal terms in the domains in which the public culture is shaped and reshaped. (Lenard 2012, 77)

No: "public cultures" will necessarily emerge from imperfect inclusion, because cultures *require* exclusive as well as inclusive formative processes. It is evidently a pipe dream that, in a pluralistic society, we can rally round anodyne sets of values upon which we all agree, such as inclusion and pluralism. We should not deny the public sphere its past; it did not emerge *ab initio* by universal agreement from behind a veil of ignorance. Its history is an essential aspect of it, necessarily failing to reflect the views of those who played no part in its construction, and necessarily influenced by values and judgments that have been challenged, rejected, or fallen out of fashion. It is a complex and perhaps sometimes embarrassing inheritance that underpins our discourse and interaction, and, while we might want to tidy it up (and it is the nature of the public realm to be always contested), it is a mistake to expect it to perform useful functions in the absence of those markers of history, geography, religion and culture that prevent it from being "as inclusive as it ought to be." The only thing it *ought* to be is present. Our interactions, with it as the medium, will do the rest.

The public sphere of a community or nation is required to support trust judgments and signals of trustworthiness (among other things). To flourish in an integrated way with the host community, outsiders need to learn how to send their own signals effectively. "But *this* is important, namely that this reaction [to training], which is our guarantee of understanding, presupposes as a surrounding particular circumstances, particular forms of life and speech" (Wittgenstein 1978, VII §47, his emphasis).

The outsider acting in good faith should gain reputation and recognition gradually via various means – observing and copying standard types of behaviour, obtaining credentials, and learning languages, customs and value systems, at least to become familiar with them, if not to share them. There is no real substitute, in putting together common views, fortalking to others and listening to them (Baier 1994d, 86–87). It is admittedly possible, as the Puritans advocate, to devise measures by which outsiders can claim a series of rights from the host *before* mastering the languages of signalling trustworthiness, but the result will only be increasing hostility and mistrust because of the manifest unfairness of such shortcuts, and difficulty in trusting reliably on the basis of this restricted palette. Many years ago, Gordon Allport argued that increased contact between groups facilitates tolerance rather than increased mistrust only if four conditions are met: equal status in the context of contact; common goals; incentives to cooperate rather than compete; and support from authorities (Allport 1954). These are certainly all desirable, but there is no shortcut to any of them. Only the fourth can be artificially engineered, but it will be counterproductive to apply it in the absence of the other three.

Does that mean that newcomers should be forced to integrate, learn the tedious rote of the citizenship test, throw away all those conspicuous foreign clothes? That we should reject policies that "permit isolation rather than demand integration, and thus fail to provide a basis on which trust can build" (Lenard 2012, 105)? No, I think not. It will depend on the community the newcomer wishes to join: *its* rules and norms are what count. There need not be a demand for, or even pressure on, a newcomer to relinquish his or her familiar ways. Cities tend to be more multicultural than smaller towns, partly because they favour shallow legibility across a broad, transient population, over rich legibility in a small but stable one.

Two vital caveats need to be added. First, there are benefits from integrating, and benefits from staying segregated, and you can't have them both *in toto*. It is a trade-off, but someone who refuses to learn, adapt to or adopt local norms won't get all the benefits of trust from the community, and shouldn't expect to (still less assert a right to them). Conversely, someone who does adapt, or trains their children to adapt, may find themselves alienated from their original community, which could be traumatic. Equally, hosts shouldn't be surprised if someone who tries hard to adopt the local norms and is still rejected or mistrusted eventually says to herself "the hell with it," and retrenches into segregation. Neither migrant nor host can have it both ways.

Second, this is also a problem we should not exaggerate – many Western democracies, such as Britain and America, are actually pretty good, taken in the round, at assimilating incomers. Many anti-immigrant demonstrations by host communities may be traced to wider problems such as socioeconomic deprivation, or cultural issues, such as a feeling of domination by remote elites. The loud volume of political debates may not reflect conditions on the ground.

To conclude, incomers surely value the communities they wish to join. A community needs loyalty to it to make it sustainable. If outsiders wish to join it, their loyalty should ordinarily precede their integration, and certainly should not be conditional on receiving the full benefits of membership. A presence in a community created and fostered by others is a first step to future benefits for the integrated outsider. In that case, it is in the outsider's interests to work to sustain the community.

It is wrong to take this as a failure of respect for the outsider; rather, it is a means of avoiding conflict and aiding integration. "The solution of the problem you see in life is a way of living which makes what is problematic disappear. … But don't we have the feeling that someone who doesn't see a problem there is blind to something important, indeed to what is most important of all?" (Wittgenstein 1998, 31).

8. Trust and the state

Conservatives can be ambivalent about the state. On the one hand, they respect authority, and as the state is the ultimate source of political authority, they want it to be strong and effective. On the other hand, they recognise its tendency to intrude into the private sphere, sometimes by its own logic and momentum, and sometimes because the hollowing out of civil society and family life leaves a vacuum that needs to be filled. They judge the state to be incompetent at dealing with private matters, and subject to capture by well-organised political actors. When it tries to help, its welfare programmes quickly turn into surveillance and intrusion (Gilliom 2001; Eubanks 2019). The key for conservatives is for the state to do the essential things capably, and not take on responsibility for the total happiness of the nation.

THE POWER OF THE STATE

"Spontaneous social action will be broken up over and over again by State intervention; no new seed will be able to fructify. Society will have to live *for* the State, man *for* the governmental machine" (Ortega y Gasset 1932, 121, his emphasis). Ortega voices a common conservative concern, but is not entirely correct, because canny interest groups learn that the way to impose their own view on society is to capture the state, to pressurise it to issue rules and regulations supporting their own ideas, values, preferences or ways of life. The groups "invest in rules and rule makers … one rule added upon another, until a simple reform becomes a thousand-page bill and sprouts twenty thousand pages of regulations" (Buckley 2018, 83).

One way of measuring the strength of the state is to look at how far it is able to resist such capture. A weak state is a battleground for groups which prefer compulsion to persuasion, be they business organisations, trade unions, professions or activists; rather than persuading voters to support their causes, they would rather lobby ministers and civil servants to get their views adopted in the corridors of power, a far easier task. Many conservatives have argued that this is a particular danger with a technocratic and exclusive elite (Schwartz 2013; Sandel 2021; Turchin 2023). The Puritans above all flourish symbiotically within and around the state.

An effective state monopolises force and exerts power legitimately. Almost all conservatives support the rule of law to constrain the state and render it predictable, but as Carl Schmitt correctly noted, it has the power to subvert the rule of law in turn. In emergencies, Schmitt argued, this power should be used to sustain the nation or political community. The rule of law is not a rule; it can't force obedience. Individual rulers make decisions, and *in extremis* they must decide how the law is applied and who applies it. As he put it, "sovereign is he who decides on the exception" (Schmitt 1988, 5). While this follows logically from the nature of the state as the monopoly source of power, so central was this insight to Schmitt's thought that it is hard to see where the stability valued by conservatives might come from. While he was a self-described conservative, his status is tendentious at best, and nowadays is valorised more on the left than the right (Laclau 2005; Mouffe 2005).

Be that as it may, if the rule of law is not to be subverted, the state must be trusted and trustworthy. Not only that: citizens also need to maintain trust in each other, because if not, there will be conflict, and Schmitt's exception may have to become the rule. As the Peasants and Puritans square up to each other, this extremely unwelcome circumstance looks ever-more plausible (for instance, after the storming of the Capitol Building in 2020 by supporters of President Trump). A well-publicised poll in 2024 suggested that 47% of Americans thought that a civil war was likely/very likely in their lifetime (MaristPoll 2024), although as ever with surveys, asking a controversial question leads respondents into unfamiliar cognitive territory.

While Schmitt's view of the powers of the state was distinguished by his ability to look unpleasant implications squarely in the eye, his view of community was inflexible and extreme. One cannot do politics or any social interaction without excluding someone, but Schmitt made this totemic: "The specific political distinction to which political actions and motives can be reduced is that between friend and enemy" (Schmitt 1996, 26).

Communities don't work like that – membership doesn't determine who one's friends are, and exclusion doesn't make one an enemy. Separate communities can mix and work together as easily as they can be in tension. They are usually porous, accepting newbies under conditions that need not be stringent. This matters, because on the Schmitt friend/enemy view, trust is fixed, a tool of conflict, rather than a means of mutual support and civility. By definition, one trusts one's friends, and mistrusts one's enemies; one respects one's commitments to one's friends, but less so to one's enemies (a Nietzschean question: can one make genuine commitments to enemies at all?). Trust will be hard to foster if the state uses its monopoly of force and law to bend rules for the benefit of friends and the disadvantage of enemies.

THE STATE AS CIVIL ASSOCIATION

An alternative conservative theory of the state was developed by Oakeshott (1975). He divided human associations into two types: *enterprise associations*, designed to do or achieve something specific, and *civil associations*, having no external purpose other than hosting its members and sustaining itself.

Reflecting on the development of the state through European history, Oakeshott conceived it as an association, although a special one. Because it has power, authority and legitimacy, it is problematic when it takes the form of an enterprise association, because it can and will compel its citizens to work towards its goals, even if they dispute them or prefer to spend their lives differently. When citizens have a standard of goodness or rectitude imposed upon them, that severs "the link between belief and conduct which constitutes [their] moral agency" (Oakeshott 1975, 158). The state as enterprise association acts as if it obediently carries out a general will – although this "will" is really manufactured by the state and bears no relation to the aggregate or individual wills of its citizens.

Oakeshott on Civil Association

On the other hand, if the state is a civil association, it holds the ring for civil society, allowing civility to flourish in a community without "taking sides" or interfering in the choices that citizens make, as long as they respect the rules of interaction. It is procedural rather than substantive, and earns authority by being properly constituted. It produces authoritative rules through an agreed procedure, but does not demand any particular purpose or form of citizens' interactions. It requires no unity among citizens beyond their mutually recognising its authority to make law, although it itself must be unified. It does not promote the interests of some individuals, or favour some ethnicities, religions or political ideologies over others. It should make virtuous behaviour possible and profitable, without defining standards of behaviour (Oakeshott 1996). This does not mean a minimal liberal state; as Oakeshott tartly observed, "a plan to resist all planning may be better than its opposite, but it belongs to the same style of politics" (Oakeshott 1991a, 26). But the state as civil association has liberal aspects, such as enacting impersonal procedures, using coercion infrequently but effectively, and acceding to its own restriction by law.

It is not the job of the state to make everyone trusting and trustworthy, but it should try to ensure that trusting and trustworthiness are possible, and ideally winning strategies. It might create or maintain institutions of global trust, for example, as part of the apparatus of civil society.

Oakeshott's conservatism was shot through with liberalism (Gray 1993a), and the state protects liberty best as a civil association. The state's provision of authoritative rules allows it to promote liberty against a clear framework of what is acceptable and expected, a background of restraint. It is not a coincidence that liberal politics emerged from, and initially became successful in, Northern European societies where the licence inherent in individualism was curbed by Protestant sobriety (Shils 1997b, 111–112, 121; Urbinati 2015, 38). Freedom demands concrete and legible order to be meaningful; without it, it is impossible to plan or achieve much, however unrestrained one is.

In such societies, smaller associations and little platoons stand between citizens and the state. These private sector enterprise associations provide niches for civil activity without restricting liberty, because their members are always free to leave them. Hence the state as civil association (a) ushers in a virtuous circle of sociability, cooperation, and virtues and habits that include trustworthiness and trust, and (b) respects the group privacy of such associations, enabling social freedom (Schoeman 1992). The state as enterprise association by contrast tends to dismantle such groups, to the extent that they do not serve its goals, and to replace them with artificial associations that do. These become large (because well-funded), but shallow (since they serve no-one's interests but the state's). In so far as they support trust at all, it is indiscriminate, on the understanding that all members are equal in the service of the state and equally deserving, undermining important mechanisms for companionship and free association characteristic of civil society.

Oakeshott denied that a legitimate state can be created by citizens' assent and social contract, but on the other hand it needs their respect for what Hobbes called "the mutuall Relation between Protection and Obedience" (1968, 728). Acknowledgment is needed of the *authority* of a pre-existing, contingent, sometimes even arbitrary (but decidedly never abstract) congeries of constraints, privileges and liberties that make up the state as civil association. This is not on contractual terms that might be withdrawn. Neither is it based on consensus about its purposes or via a general will, because that will erase plurality and create an enterprise association. Citizens need not even approve of the state, as long as they recognise its authority to set the rules of civil engagement, and their obligation to obey the law (Gerencser 2000, 125–126).

Civil society, in a strong sense, is just the joint acknowledgment of the state's authority; hence the *demos*, while plural, is unified in its recognition of the state. Oakeshott's vision strongly supports Burkean representation from the citizen's perspective, complementing Burke's account of the representative's view. "To authorize a representative to make a choice for me does not destroy or compromise my individuality; there is no confusion of wills, so long as it is understood that my will is in the authorization of the representative and that the choice he makes is not mine, but his on my behalf" (Oakeshott 1991c, 281–282).

Oakeshott was impressed by Hobbes' argument that the state makes itself visible via a sovereign (or sovereign body) that is clear, phenomenologically evident, and legible, whom citizens acknowledge has the authority to obligate them to keep to its rules (Oakeshott 1991c, 260, 277). Because Oakeshott's state is voluntarist, this sovereign, unlike Schmitt's, cannot decide on the exception. The state's authority, once established, is powerful, but cannot make demands on citizens' thoughts; it can only regulate their actions in so far as they affect the conduct of the public sphere. Hence Oakeshott's interest in Hobbes, who "is not an absolutist precisely because he is an authoritarian" (1991c, 282). For Oakeshott, Hobbes was a sceptic who nevertheless avoided the pitfall of nihilism, and created an account of authority (Gerencser 2000, 119–121), ticking a number of conservative boxes, including scepticism about rationalism, and the need for legibility.

Filling in the Gaps in Oakeshott's Account

Oakeshott's view is attractive in many ways, but is ultimately inconclusive.[1] The distinction between civil association and enterprise association is hard to make out at the limit, and it may be a continuum; the formal rules of civil associations may with sufficient ingenuity be rewritten as substantive principles typical of enterprise associations, and *vice versa* (Gerencser 2000, 133–134, and for a real-world example, Legutko 2016, 160–164). Oakeshott himself believed that the modern state is a contingent and complex mix of both civil and enterprise associations that contradict and impede each other, "two powerful and contrary dispositions, neither strong enough to defeat or put to flight the other" (Oakeshott 1975, 323), so no state is a "pure" version of either.

In a civil association, associates are only required to modify or qualify their actions according to the rules while pursuing their self-chosen ends; they are "related solely in terms of their common recognition of the rules which constitute a practice of civility" (Oakeshott 1975, 128). Yet that in itself is a substantial connection, and the formal rules of the civil association may blend with the substantive norms of the society, both being absorbed into practice. Hence, realistically, the "self-chosen" ends are selected from a palette limited by a context set by rules, traditions and forms of life – aspiring to get a job, get married, go to university, get a car, a house, a mortgage, a pension, an annual

[1] It may also need to be updated given the evolution of the state in digital modernity. I hint at this topic in the concluding chapter, but full discussion must await a different venue.

holiday. And these are connected to various important practices and institutions, such as – in Europe – Christianity, which was central in creating and structuring those forms of life (Spencer 2016). While the important relation between members of the community is that they all recognise the authority of the state or sovereign, they will relate in many more substantive areas as well, providing a shared understanding of context which then, among other things, supports mutual legibility, trust judgments and trustworthiness signalling.

This is great for trust, but undermines the radical plurality which interested Oakeshott. Ultimately, the unresolved issue is Oakeshott's agreement with Hobbes that authority grows out of citizens' recognition (or, in Hobbes' word, consent). Where does this unity, even the minimal unity of the civil association, come from? Even if citizens only have to agree on the sovereign's authority, that is still a big thing to accede to. And conversely it is not that simple for someone to decide that the law doesn't have authority and therefore doesn't obligate them. Oakeshott criticised rationalists for trying to engineer incoherence out of society. Was he, in trying to engineer incoherence out of the foundations of the state, vulnerable to his own critique?

There are two directions that lead from this thought. One is that the Oakeshottian account can be made more radical, allowing citizens to leave the civil association state, deny its authority, and indulge in acts of resistance such as civil disobedience (Gerencser 2000, 153–165). This is a view characteristic of liberalism, which will set problems for trust (and will be revisited in Chapter 10).

The other, which I would advocate, is to concede that the decision to acknowledge the authority of the sovereign is not taken by autonomous individuals, but rather within a matrix of standard practices (family, marriage, education, capitalism etc.), and belief systems (e.g. Christianity and liberalism, in the West), following training and socialisation. An individual agent, for Oakeshott, is grounded both in a personal history and in the history of his or her community, enjoying "independence ... in respect of being a reflective consciousness composed of acquired feelings, ... understandings, beliefs, convictions, aspirations, ambitions, etc., recognitions of himself and of the world of *pragmata* he inhabits, which he has turned into wishes, and wishes he has specified in choices of actions" (Oakeshott 1975, 40).

In other words, by the time Oakeshott's agents, with their reflective consciousness and the *pragmata* of the world they inhabit, have developed agency, they have already absorbed a form of life that is the civil association, which includes within it a number of basic norms of trustworthiness, one of which is acknowledging the authority of the sovereign and the law. Even though Oakeshott wanted his society to be maximally liberal, it holds together only when the community/association can exert centripetal force through socialisation, so that trust and trustworthiness can rest on mutual legibility. The importance of the sovereign now emerges; as it is evident and legible to citizens, it can be the object of relatively accurate trust judgments. What Oakeshott characterised abstractly as mutual acknowledgment of the state's authority is better seen as citizens' trust in the sovereign, which is more often learned and implicit in their behaviour, than explicitly placed by free decisions of autonomous agents. They don't *will* their recognition of authority, they *learn* it at their parents' knees.

The state is sustained by, and is there to sustain, a culture within which deep and rich trust can be nurtured. The cultural dimension is important: we need to resist the idea that common rules can replace common culture (Gray 1995, 112–117). Religious freedom does not mean that no religion can be accorded a privileged position because of its historical importance, and similarly for political freedom. Oakeshott himself emphasised that a culture that can sustain a society with a collective commitment to civility and a practical sense of how to implement it in everyday practice is itself an achievement (perhaps an historical accident that can't easily be replicated).

THE LEGIBILITY OF THE STATE AND THE INTERFACE WITH THE CITIZEN

Where the state has long been a structuring presence in a society, the removal or reduction of its presence can *reduce* legibility. Attempts by neoliberal authorities to minimise the public sector may create space for private actors with theoretical powers in law to protect their interests, but that does not give them understanding of how to do this in a practical sense. This is one way of understanding the absence of civil society in Eastern Europe following the collapse of the Soviet Union (Krastev & Holmes 2019).

More usually, however, the state's presence grows, especially in its guise as an enterprise association. As it renders its citizens legible to itself in order to deploy their powers for its goals, it simultaneously renders civil society less legible to its members (Scott 1998, 191), usually without making itself more legible in compensation. It manipulates and judges citizens' trustworthiness to it, while citizens' resulting alienation degrades their trust judgments of both the state and each other.

> [People] abridge their own practice in order to be legible by the state – for instance, an unemployed person on welfare might be better off working casually in the informal economy, but the state recognises only the possibility of formal employment or enforced idleness. … If she forfeits her welfare entitlement, the social safety net is removed from under her, but if she claims welfare and foregoes informal employment she is unable to use her contacts and local knowledge (her social capital) to help support her and her dependents, and work that would benefit the local community is left undone. (Coles-Kemp et al. 2018, 46)

The legibility of the state is conditioned in the real world by interactions between citizens and its functionaries; not so much the senior figures such as judges and policymakers, but those public-facing officials who deal with the citizenry on a daily basis. They explain processes to citizens, implement higher-level policies, and ultimately are gatekeepers who decide the official status of a citizen's problem. They are fellow community members, but they determine whether their relationship is transactional, respectful, or humiliating (Zacka 2017).

They are also vital for the state. Their decision-making makes vaguely written regulations concrete. They prioritise the range of typically conflicting imperatives that trickle down from the mandarins above them (e.g. to save money, improve security, or target particular social groups). They make the system work even when it is under-resourced (Zacka 2017, 40–41). They have a choice between being trustworthy for their managers, or for their clients in the citizenry (Zacka 2017, 62).

To offset power asymmetries, trustors and trustees negotiate the nature and context of the latter's commitments. In this case, while bureaucrats have the upper hand and make the final decisions, clients of the state can also influence outcomes (Johannessen 2019). They can be smart or scruffy, hostile or courteous, can ask advice or present demands, drag their children to meetings, and so on. While the bureaucratic environment tries to make the interaction routine, the human element on both sides of the desk (or phone) will necessarily intrude.

To regularise these complex interactions, it is often thought sensible to reduce bureaucrats' discretion, either to treat them as mechanisms, or to introduce the literal mechanisms of digital government. In other words, to implement trustless trust.

Trustless trust is often seen as a liberal *desideratum*. The rule of law, it is maintained, will be better preserved if decisions are made by elected politicians or accountable officials. In the ideal, "the basic trust that extends among citizens and political actors is so strong that the likelihood of violating it appears impossible and therefore as choiceless. The ongoing practice of effective democratic politics relies on this 'choiceless' trust as a matter of course" (Lenard 2012, 43). Thus, the fiction of the social contract can be preserved; voters consent to the procedures and decisions that affect them, and politicians consent to their restraints.

In particular, digital government delivery systems withdraw the state from in-person contact, and impose trustless trust both on the suppliers of state services and the demand side. Citizens struggle to fill in the online forms when their circumstances are unusual, and substitute workarounds where their answers are problematic. In the absence of a sympathetic human interlocutor, the interaction becomes antagonistic, and the state looks less like a safety net and more like a nut to be cracked.

There are at least three problems with converting government bureaucracy to trustless trust. First, the exact aims and scope of policy will need to be formalised and set out unambiguously. That is easier said than done in a complex system where legislation is drafted by civil servants, influenced by ministers, specialists and lobbyists, makes its way through a legislature, is amended in various ways, is implemented in digital or bureaucratic procedures, and ultimately meets up with the courts to be judged alongside other legislation and constitutional rights.

Second, substituting a technical procedure for one mediated by humans won't leave it unchanged. Negotiation is eliminated. Humans are generally visible and accountable, whereas digital systems are created by small expert cliques accountable only to the commissioners of the software. The communicative aspect of the interaction is replaced by an interface where the claimant identifies herself in terms meaningless to her, that the state recognises (e.g. a biometric or a password). This is not merely a change in interaction style but a removal of fundamental inputs to a trusting relationship between state and citizen (Coles-Kemp et al. 2018, 45–46).

Third, the interface with the citizen can be captured by spurious "stakeholder groups" that thrive on a deep relationship with officials, so that a supposed consultation between government and citizen becomes a dialogue between the managerial cadre of civil servants and special interest groups. Mutual trust flourishes between them; indeed, bureaucrats themselves are a special interest group (Niskanen 1994) with a number of organisational tactics at their command, such as which experts to invite to meetings, and how they summarise a discussion for their superiors (for an egregious example, Sullivan 2021). Trustless trust merely cements this cartel in place.

Trustless trust can be imposed for a number of reasons, and where digitised may deliver policy efficiently and cheaply, create lots of data for the government, and standardise the citizen experience. Where the bureaucracy is corrupt or inefficient (and trust therefore low), as in India, digital government has plenty of upsides (Naidu 2000; O'Hara & Stevens 2006, 152–166; Aiyar 2017; Sharma 2020; O'Hara & Hall 2021, 183–199). However, citizen trust in government will depend less on a learned and informed attitude (since citizens will be less informed), and more on trust of the Internet and technology generally, including the usability and speed of the interface, and perceptions of system security and privacy (Janssen et al. 2018). This will fundamentally change the nature of the relationship between state and citizen, and Hobbesian and Oakeshottian theories of citizen acknowledgment of state authority are likely to look increasingly anachronistic.

PART III

Blockchaining democracy

Introduction

The preservation of trust, or rather conditions conducive to it, is a central conservative cause, because trust and trustworthiness are important virtues, essential to the maintenance of little platoons, beneficial communities, and the smooth running of the state. Trying to restrain human behaviour by mechanisms and reducing discretion and freedom with trustless trust cannot banish trust, as we still have to trust the systems that implement it (Janssen et al. 2018). The blockchain-inspired idea of setting up a system so that *no person* is trusted, *only* the system, increasingly characteristic of today's politics, should be rejected, because trust is an essential part of sociality.

To begin Part III, we return to the three tenets of trustless trust from Chapter 1, and review the conservative response that emerged in Part II. First tenet: *centralised institutions or intermediaries can never be assumed to be trustworthy. Centralisation is inefficient and creates opportunities for domination – control the centre and you have power. Disintermediation enables localisation and individualisation.*

Conservative response: This group of claims is ideological, not empirical. Hierarchies often function well, supported by trust between subordinates and superordinates. In a decentralised system, it is hard to achieve particular goals or follow a strategy, whereas a centralised system can be steered by those at the centre, and a hierarchy by those at the top. Decentralisation *is* helpful for allowing a network to increase in scale, but that means a trade-off between system growth and strategic deployment; where a system serves an important purpose, the pursuit of growth can undermine it. Power can certainly be misused, but there is nothing necessarily bad about power or power asymmetries, which are essential in many relationships, especially ones that are crucial to socialisation, such as parent-child, clergy-congregation, doctor-patient and teacher-student. The ultimate hierarchy, the state, depends on trust and (when it functions well) fosters it too. Trustless trust proceeds from the instinct to be independent and beholden to no-one, and technology has enabled strong versions of that to be realised (Deneen 2019, 104). However, as the mechanism has to be trusted at a minimum, such independence is a myth.

Second tenet: *people can never afford to trust each other. Trust is a type of coercion, and those in competition with each other will always do each other down.*

Conservative response: The reflections of Parts I and II show this critique to be mere cynicism. Abusive relationships can rest on a trustee's oppression by a trustor, but to suggest that human relationships are generally like this is unjustified. Baier at one point seemed to claim that heterosexual marriage is characterised by abuse (1994e, 120–123), but if so, it would have to be explained why women remain keen on marriage, especially in the wealthier strata of society, and why parental marriage has such a positive effect on the life-prospects of children (Kearney 2023). Similarly, many argue that coercion is at the back of capitalism (Baier 1994f, 131; Pezzarossi 2019; Cohn 2021), their knee-jerk anti-business stance belying obvious evidence to the contrary, such as the greater success of capitalism in high-trust societies (Fukuyama 1995), and the greater and more egalitarian prosperity in capitalist ones (Zitelmann 2019). And all may be fair in love and war, but competitors often place trust in each other, from poker players to, as we shall see in Chapter 9, Bitcoin miners.

Third tenet: *decentralisation, mechanism and transparency are the only reliable means of supplying trust between strangers.*

Conservative response: This again is belied by Part II, where we saw how the legibility of communities, the trust relations within them, and the basic functions of the state, support and are supported by trust in a virtuous circle. People aren't exclusively out for number one; most extend more of a hand to those they encounter than abstract models of rational economic people would suggest. The untrustworthy are rarely popular. Communities provide useful support systems and knowledge relevant to trust and trustworthiness, and these, far from being supported, are threatened by modern attempts to make trust more inclusive and less partial.

If trust is a vulnerability, it is a strange kind of one. We are technically vulnerable in all sorts of ways, from getting on a bus with a load of strangers, to putting our tenderest feelings at another's mercy, to investing our wealth. Any such transaction can go wrong, and some do, but Western societies have evolved to manage the vulnerabilities, to all our benefits. Only novelists and film-makers revel in dark narratives of murder, misogyny, hate and moral ambiguity under the apparently bland social surface. If entertaining, their work is absurdly unrealistic – in real life the bland surface usually conceals only further blandness, amusingly expressed in a memoir of British suburbia entitled *Where Did It All Go Right?* (Collins 2003).

To claim that trust is primarily a vulnerability is to ignore both the joy of a trusting relationship, and the terrible pathology of being unable to trust. Conservatives use the matrix of a legible community to stifle theoretical concern about the uncertainty of trust, so that one can aspire "to be equal to one's own fortune, to live at the level of one's own means, to be content with the want of greater perfection which belongs alike to oneself and one's circumstances" (Oakeshott 1991d, 409).

The final task of this book is the negative one of examining ideologies that are antithetical to trust, either because they explicitly reject trust as a vulnerability and prefer to replace it with trustless trust, or because their analyses of trust are so flawed that the measures they advocate are counterproductive. Chapter 9 looks at technology and technocracy (apolitical government through technical measures), an ideal for the rational and efficient management of messy society for a century and more (Oakeshott 1991a), and the driving metaphor behind processes of modernisation. Chapter 10 considers the dominant Western ideology of liberalism, Chapter 11, critical accounts from the post-Marxist left, and Chapter 12, identity politics.

Much of course has been written on these ideologies. My aim here is not to rehash old debates, but rather to highlight those aspects that impact directly on social trust.

9. Technology against trust

Technologists have always been interested in improving human production and communication. The Luddites complained about machinery putting skilled textile workers out of work in the 1810s, but the net result of the new technology was cheaper clothes (possibly of inferior standard) available for many more people. Similar arguments play out nowadays about artificial intelligence (Cowen 2013). The major proponents of digital modernity believe that the ambient data infrastructure (see Chapter 1), with its focus on correlations rather than causes, honed by Silicon Valley methods, will solve pretty well any problem, using representations and procedures for probabilistic inference at scale, and managing tradeoffs between precision and timeliness of inference to maximise the expected utility of outputs (Gershman et al. 2015). Crunching data gathered by the ambient infrastructure with AI enables the manipulation of individuals, and the physical and virtual environments that they inhabit, leading, it is hoped, to more rational and pro-social behaviour in the interests of the individuals themselves (Altman 2021). Patrick Collison (a self-made billionaire) and economist Tyler Cowen have advocated a new discipline called "progress studies," to explore how self-made tech billionaires and economists can improve the world (Collison & Cowen 2019).

Technological innovation is increasingly regarded as necessarily *disruptive*. If an invention *doesn't* put someone out of business, it isn't doing its job (Christensen 1997; O'Hara 2022a, 65–77), extending Schumpeter's famous characterisation of capitalism as "creative destruction" (1950). One influential paper argued that scientific *progress* was slowing down because it was getting *less* disruptive, as measured by patterns of citation of previous work (Park et al. 2023).

Disruptive innovation is problematic, because it threatens contextual integrity. Different people or groups may lack the concepts to describe the new experience, or their concepts may be misaligned with it, so that communication and discussion are rendered difficult, and they will struggle to adapt the old norms to the new environment. It upends signalling systems, making it harder to assess trustworthiness.

The aim of trustless trust is the ultimate disruption: of trust itself, undermining "trust in trust" (Gambetta 1988; Luhmann 2017, 73–78), and using technological contrivance to implement mechanism solutions. I will consider this direct type of disruption in the opening section of this chapter. Indirectly, the development of technologies such as Bitcoin and the blockchain is intended to disrupt common bureaucratic processes, removing their uncertainties and indeterminacies. This kind of disruption is explored in the second section.

DISRUPTION AND DISINTERMEDIATION

Disintermediation is a common argument for technological improvement, because who could object to cutting out the middleman? A process to be disintermediated is conceptualised in technological terms of an input and a related output (e.g. it matches a lively gentleman GSOH directly with a hopeless romantic and dog lover). What happens between input and output could be a black box, as what goes on between them is irrelevant to the function computed. It matters not whether it is a human broker or a computer program.

Yet though the black box is irrelevant to the *technology*, it is not irrelevant to *society*. The black box now conceals the locations where the users of the service or process come to understand it, to welcome or critique it, and – ultimately – to trust it or not. The existence of an agent working in your interests within the system is worth paying a percentage for. Even if the agent is also working for the other side of the transaction as well, everyone benefits from a successful conclusion.

The Joys of Disintermediation

When a form of life is disintermediated, the trust it engenders suffers. There is almost always a financial gain for someone – often including the customer who pays less, or even zero, for a service. This is generally enough to render the intermediary otiose, but at a social cost of trust and contextual integrity. Sometimes this will be worth it, sometimes not. We can look at three types of occasion where the gain can be challenged.

First, a platform such as Amazon or eBay can make matches, for example of buyers and sellers, efficiently and cheaply at scale. However, what is called "disintermediation" here is really a new kind of intermediary, an automatic broker of access and communication. The new automated platforms can build up such large networks as to dwarf the old intermediaries' market-making power (Knee 2021). Many think their fees and power have grown out of all proportion to the services they provide; for example, Amazon has been accused of exploiting its dominance in various ways, including anti-competitive practice, poor treatment of its workers, and tax avoidance. Sometimes, platforms use economies of scale to charge no fees at all, instead charging rent in monetisable commodities, most obviously personal data (Zuboff 2019).

Secondly, once trustworthy human intermediaries go out of business, the services they used to provide disappear. As one example, digital travel services have more or less replaced the friendly travel agent who used to organise your holiday. But online travel agents such as opodo.com are information aggregators with little expertise, and less customer service – disastrous when you are stranded somewhere and need help in a hurry. You could trust the travel agent, and reach them to discuss matters if they let you down; you cannot trust opodo.com.[1] As companies try to save money by disintermediating their own customer services, replacing them with bots and FAQs, customers are made to look up the solution to their own problem themselves. The universal result is the decline of customer satisfaction, as we all try to work out how to get through to a human.

Thirdly, the technology is only as good as its data, so it had better be genuine. It did not take long for buyers and sellers to learn how to game recommenders such as Amazon and Tripadvisor with spurious reviews, and then to supercharge their efforts with bots, and ultimately with generative AI systems like ChatGPT. It has reached the point where only black box AI systems can detect fake reviews well (Salminen et al. 2022). How confidently can we place our trust in the midst of such an arms race?

That is not to say that digital intermediaries are always bad. In low-trust environments, online platforms have become successful precisely *because* they provided a scalable global trust mechanism, to reassure buyers and sellers. Alibaba played this role to help develop e-commerce in China, and sites like Kavak, a used car marketplace in the Americas, Vendorstack, a service provider for small businesses in Nigeria, and Tendo, a Ghanaian platform along eBay lines, are doing something similar in their own challenging business environments. The problem is that technological advances sweep away both useful and redundant intermediaries without discrimination.

Information Intermediaries

The disintermediation of most public concern has been that of news and information provision, where large-scale digital aggregators and advertisers have undermined the business models of old-style media. The Internet, by providing information free of charge, disintermediated the gatekeepers that we used to call journalists, editors, experts and librarians. That was liberating – instead of a staid, fixed, establishment view, we got competing positions, heterodox

[1] I declare an interest. A stupid error on the part of opodo.com once cost me thousands of euros, with no customer service response and no ability to put it right. I am not alone (Baker 2020). "Opodo could keep my column going all by itself given the regularity of complaints I receive" (Tims 2023).

voices and the invigorating clash of debate. Information could be sent to you depending on *your* interests and social networks, not on *others'* judgment of its significance.

Unfortunately, three related problems have made themselves felt. First, the fixed establishment view was fixed for a reason – it had emerged from investigative and ethical processes – methodical journalism, statistics and science – that were likely, if not guaranteed, to produce trustworthy information. It wasn't a perfect system, everyone has an agenda, and the establishment has been known to cover up mistakes and bias. On the other hand, the opinions of influencers and activists are divisive, comforting (to followers), infuriating (to opponents) and misleading (to everyone).

Misguided attempts by sociologists, particularly in the field of science and technology studies (STS – Sismondo 2010), and in the strong programme in the sociology of science (Barnes & Bloor 1982), to recast important notions such as scientific method, rationality and reason as purely exercises of power by dominant groups have been very influential in reducing trust in scientists, journalists and other gatekeepers of knowledge. These academics of the left, having spent decades undermining epistemic authority and trust in science, are now wondering why conspiracy theories (Cassam 2019; Rothschild 2021) and denial of climate change (Kahn-Harris 2018) have such currency among the Peasants and populists of the right. While the original insight that science's social aspects were relevant to its conduct and outcomes was valid (Kuhn 1970), its implausible extension by STS researchers to the idea that science is *only* social and *only* an exercise of power is perhaps the most spectacular own goal in academic history. Conspiracy theorists use the very critiques championed by STS to debunk the science that shows them to be wrong. Thus, once more the Puritans enable the Peasants.

Secondly, whether or not influencers and activists are sincere, their prejudices and those of their supporters can be exaggerated by technology which automatically invents stories that appeal to them, coordinates armies of trolls to provide partisan or divisive content, or uses algorithms to send the right catnip to the right cat (Pomerantsev 2019; Howard 2020; Ruffo et al, 2023). People receive the stories and posts that they will react to most strongly, and get a personalised and biased view of the world. This might be malicious, as when botnets spread misinformation and conflict, but it is also an "innocent" by-product of many social networks' business models, where more clicks lead to greater advertising revenue.

Thirdly, now the Internet is the default information space, it has become the leading source not only of news, but also evidence. The use of deepfake technology to produce convincing and deliberately misleading video and audio is, despite some concern, still in its infancy, but it will soon be possible to show political figures saying appalling things, soldiers committing atrocities, or identifiable people taking part in sexual acts, that are entirely fictional (Meikle 2023). The mere existence of deepfakes can also provide a defence for people who have been caught on film doing wrong, who can now falsely claim they are a victim of deepfake fraud (much as President Trump got into the habit of dismissing criticisms of his behaviour as fake news). Less dramatically, data can be faked and placed online, perhaps to support the conclusions of a fraudulent scientific paper, or for sabotage, as in the technique of "poisoning" the training data of machine learning algorithms (Tian et al. 2022).

The upshot of these three issues has been a fragmentation of information resources across media and on ideological lines. Hence the production of a common view that could underpin basic agreement and consensus across a community has become so much harder (Lenard 2012, 168n.12). There is probably no way back to the old days of the BBC and quality broadsheets, but we clearly have a technology-based trust problem. Trustworthy intermediaries can no longer support the expensive processes of checking and sourcing information that make them trustworthy, while too many consumers of information prefer patently untrustworthy sources.

When information is disseminated by social media, its *bona fides* rest on being generated by the audience's network. The trustworthiness of the source, from the perspective of the audience, is evidenced not by a trustworthy intermediary, but the joint membership of source and reader of a network. Ideally, trust should develop under the aegis of a form of life conducive to social trust, but social media sharing and matching is merely a degenerate, decadent form of narcissism.

We are in a tricky new information space, and we need to learn how to navigate it. This is not, in my view, an area in which we can rely on the state to act as information intermediary of last resort (for an opposite view, Sunstein 2021). Neither do I think regulating social media or using automated fact-checkers and transparency will necessarily work (for opposite views, Pomerantsev 2019, 229–240; Howard 2020, 137–169). The solution cannot be to empower people to do something (i.e. constantly evaluate the information they receive) that they are not particularly inclined to do. We should not delude ourselves that misinformation is a problem only of supply. There is plenty of demand, not least from the Puritans and the Peasants, which incentivises the supply; hence regulating supply will not do the trick for the same reason that the war on drugs failed to stamp out drug-taking.

Trustworthy Supply of Information

How can information suppliers be as trustworthy as possible to the widest range of people? Trustworthy information suppliers go beyond their commitment to the truth. A conspiracy theorist who believes that the world is being run by shape-shifting lizards is no doubt sincere, and wishes to persuade others from the best of motives, but – since the world is patently not so run – she is hardly a trustworthy source. A rough set of guidelines setting out the commitments an information provider should make to a wide audience should include:

(i) Provide information even if it might be disappointing or disillusioning to the supplier, and encouraging to opponents.
(ii) Search for reliable evidence from dependable sources and, where a source is not known to be trustworthy, test its claims.
(iii) Be sceptical (i.e. not automatically trusting) of all information, even that which it favours.
(iv) Be properly respectful, if not deferential, to relevant acknowledged expertise.
(v) Try to avoid vocabulary or presentational effects that might sway the reader, or provoke particular positive or negative reactions, without a justification in terms of content.
(vi) Be respectful of the possibility of alternative views, without necessarily giving them credence.
(vii) Do not try to micromanage the reader's response.
(viii) Nurture and protect a reputation across partisan divides for trustworthy reporting, based on the highest standards of public and collective epistemology.
(ix) Be prepared to acknowledge fallibility, by following up criticisms, publishing admissions of error, or explaining conclusions in more detail, where feasible and reasonable.

These nine measures would have to be supported by a business model (and compete with all sorts of clickbait), but in aggregate they set out an ideal of trustworthiness. They also demand managerial intervention – they are not amenable to automation, so that the human editor cannot be disintermediated.

If the provider is subsidised by the state, it may be hard for it to resist a political steer. If it is independently financed, then it is likely that providers for business consumers will be the most trustworthy, since investors risk losing money if they are misinformed. They care! Providers for the general public may have to compete with other forms of diversion, and so may be more likely to judge information on its entertainment value. Ideologically-slanted providers may provide biased news, as readers/viewers are not materially affected by false or misleading stories. But even where the provider is biased because funded by the state or ideologically partisan, it can still be trustworthy according to principles i–ix.

Trustworthy providers will sustain and signal trustworthiness to a wide audience, rather than a narrow, partisan group, but the trustworthiness principles neither assume a neutral or disinterested view, nor preclude attempting to persuade the reader of the merits of a particular view. Nothing about the nine commitments rules out a partisan channel. There is a difference between those who wish to *persuade* readers using reason and argument, and those who wish to *provoke* readers using the techniques of advertising, cognitive psychology, public relations, rhetoric, sophistry, manipulation and, at worst, misinformation.

The situation certainly shouldn't be exaggerated to support censorship or regulation of news media. For example, many, especially the young, seem to trust information from social media almost as much as they do professional news organisations (Liedke & Gottfried 2022). This is an attitude that they need to grow out of, and probably will, but it may not matter very much if they don't. It may be that those who now get their news from a cartoon fish on TikTok[2] got no news at all when it had to be sought out in newspapers or at set times on broadcast TV. Was that better or worse than our current situation? Probably worse.

Returning to the pre-Internet days is not a runner, but the partisanship, mistrust and Manichaeanism of politics, journalism and academia is the part of the current malaise that can probably be most easily jettisoned by audiences, if not by successfully managing our anger, at least by rediscovering the value provided by information intermediaries. This may already be happening – as social media companies have pivoted to entertainment, the proportion of people who read news on them has been falling in the mid-2020s. Many users are migrating from open networks with their clickbait to closed groups on messaging apps such as Telegram and WhatsApp. These are unmoderated, but at least the presentation of news need not be influenced by sensationalism.

We generally respond to crises in trust neither by carrying on as usual, nor by ceasing to trust, but by reorientating to new circumstances, and positing (however hopefully) a new, stable normal. It is quite possible that as awareness of misinformation spreads, this is happening in the information space.

TRUSTLESS TRUST: THE DOOMSDAY MACHINE AND THE BLOCKCHAIN

The second type of technological disruption harmful to trust is the technological performance of trustless trust without human intervention, of which blockchain is the paradigm. However, trustless trust doesn't take place in a vacuum

2 I kid you not. https://www.tiktok.com/@therealisticfishhead?lang=en.

– it has a social context which can affect the outcome drastically. We begin by illustrating its logic, and shortcomings, with a tongue-in-cheek example from the Cold War.

Dr Strangelove, or: How I Learned to Stop Worrying and Love Trustless Trust

I will follow Brian Skyrms in illustrating the logic of trustless trust with the Doomsday Machine from Stanley Kubrick's 1964 film *Dr Strangelove*. This is a fictional device that the Soviet Union has built that will, in the event of a nuclear attack from the United States, retaliate with world-destroying force. It is tamper-proof, and cannot be turned off. The removal of uncertainty changes the game-theoretic calculations of both sides. Why? What problem was it built to solve?

Suppose there were no Doomsday Machine. If the Americans made the first strike, there would be harm to the Soviets. If they then retaliated, harming the Americans, the Americans would strike again, causing further harm, and so on. But if the Soviets did not retaliate to the first strike, they would suffer less harm (just that of the first strike). Hence harm to the *Soviets* is minimised if they do *not* respond, and a rational leader would take this into account. This reasoning is available to both sides, so the Americans are aware of Soviet vulnerability on this point.

If the Soviets built a Doomsday Machine that they could turn off, retaliation is still a losing strategy. "For surely if there were an attack, it would be better for the USSR to suffer the effects of the attack than to suffer the combined effects of the attack and the doomsday machine. After an attack, if they could, they would disable the doomsday machine. And if their enemies could anticipate this, the doomsday machine would lose its power to deter aggression" (Skyrms 1996, 22).

Hence "credible implementation requires removing the possibility of choice" (Skyrms 1996, 41). The Machine is tamper-proof, not to protect it from the enemy, but to prevent its builders from having second thoughts, just as Odysseus got himself strapped to the mast to listen to the sirens.

The logical difficulty of how to make commitments binding is the core of trustless trust theory, to try to reduce the number of active choice points as close as possible to zero. Even a gun to the head can be faced; many noble souls have been tortured to death rather than betray colleagues, religion or country. But this only works if the opponent knows about the threat The *Soviet* Doomsday Machine must be present in the *Americans'* calculations. No point keeping it secret. Unfortunately, in the film, the Soviets have neglected to tell the Americans about it, and so they haven't taken its logic into account, rendering it useless.

But that is not what goes wrong (*contra* both Skyrms and the American President in the film). In a social world, logic will get you so far, but contingencies rule. In the film, a mad American officer triggers a crisis that is beyond the control of the Americans, and hence *even had the existence of the Machine been known*, the scenario would still have unrolled as it did, though perhaps less amusingly.

In the real world, deterrence worked via a series of agreements between the Soviet Union and the United States, which between them sustained enough trust to avert war. The warrant for their minimal but mutual trust rested on the very simple game-theoretic scenario of mutually assured destruction, with just two significant actors. Even that logic has now broken down, unfit for purpose in a multipolar world with three nuclear superpowers, two of which may or may not cooperate. American leadership is no longer trustworthy, with Trump's Republicans able to sow doubt about the constancy of its commitments. Russian leadership, unlike the Soviets', is untrustworthy and non-rational. Various unstable powers are also developing weapons and have their own agendas.

We can't restore the old bilateral logic, but trustless trust isn't going to help (even if it has advocates). The modern equivalent of the Doomsday Machine is the powerful technology of generative AI, which could produce its own policies on the fly in milliseconds. The problem is that its outputs are inherently unpredictable, unlike the Doomsday Machine, taking into account millions of scenarios based on its own current models of the strategies of the other nuclear-armed nations, raising the concern that "a new generation of artificial intelligence (AI) enhanced cyber capabilities will amplify the risk of inadvertent escalation caused by the co-mingling of nuclear and strategic non-nuclear weapons and the increasing speed of warfare, thereby increasing the risk of nuclear confrontation" (Johnson 2019, 442).

This takes us to a distressing conclusion. Because no nuclear power can trust that the others will not rely on AI, trustless trust may therefore be inevitable in a reckless race to the bottom: so fast is the technology that putting a human in the loop puts you at a real disadvantage against those who automate irresponsibly.

Blockchain

The term "trustless trust" was originally coined in honour of the blockchain. Blockchains are the ultimate, because no-one is in charge of them. That means no-one has individual responsibility for decisions, which, for example, may be problematic when we try to regulate their processing of personal data (Ibáñez et al. 2018). What is the basis for preferring to trust them instead of people?

Certainly not their spivvish record. The mind boggles at the number of frauds and hacks over the years, from spurious investments in new cryptocurrencies or non-fungible tokens (Tiwari et al. 2020; Hornuf et al. 2022; Mazorra et al. 2022), to insecure exchanges (Trautman 2014) and fake websites (Bhushan et al. 2017).[3] The high value of Bitcoin has long been put down to investors either seeking risk in a low-interest world, or hedging other investments, or merely following the crowd. An alternative explanation, hardly mentioned, is that investors expect cybercrime, sanctions busting and money laundering to increase, and see Bitcoin as the likely beneficiary (Trozze et al. 2022; White 2024).

Trust in blockchain is usually linked to its structure, including the transparency of the chain (it can examined at any time by anybody) and its integrity (it cannot be altered). As already noted, this last is not entirely true. It can be altered, if enough members of the network agree. This is called a *51% attack* (although it would probably need more than a bare majority).

The blockchain sustains itself by some members of the network (*miners*) proposing additional blocks to add to the chain. Each new block links back to the existing chain and adds new information; only one block is added at a time. Each miner designs a block that records relevant changes in the underlying ledger, and competes to add it to the chain. When information in a block is added, that makes it "official" and permanent. When Alice pays Bob a quantity of bitcoin, news of the transaction is spread informally around the network. Eventually, a miner puts the information into a block and successfully adds it to the chain (Alice may have to leave a "tip" to give him the incentive to do that). At that point, the transaction is recorded. Because the record is complete and definitive of Bitcoin, it can easily be checked that Alice has enough in her wallet to cover the payment, and that she hasn't used the money to pay two people simultaneously (O'Hara 2015).

To alter such a record, one would have to unravel the blockchain back to the point at which the information to alter first appears. If Alice wanted to revoke her payment, she would have to unpick the blockchain by removing each block until she got down to the one where her payment is recorded, remove it, replace it with a falsified one, and then build the chain up again from that point. Where there are thousands of members, each storing their own record of the chain, that would be virtually impossible. The integrity of the chain and the blocks that make it up are secured by the vigilance of the members. But the problem remains even if Alice can assemble enough of the network for a 51% attack. In the ideal network of thousands of independent individuals, this would be

[3] See https://cryptochainuni.com/scam-list/ for a regularly updated list of fake exchanges.

hard – she would still have to solicit and incentivise a majority of them *without* alerting the minority.

However, this idealisation doesn't fit reality, because blockchain miners do not act as individuals. The miners create blocks with complex computational hardware; successful miners are rewarded, but it needs a big investment up front and no guarantee of success. The competition between them to mine the next block can be managed in different ways, but they tend to involve either fair competition (proof-of-work, very expensive) or a measure of seniority (proof-of-stake, which has the effect of making some miners more influential than others). So competitive has proof-of-work become that state of the art semiconductors need to be assembled, powered by a lot of electricity. Miners therefore join together to create *mining pools* to cooperate and share rewards. Either way, some miners are more equal than others. What really counts is not the support of 51% of the network, but control of 51% of the computational or voting power in the chain.

Hence getting the cooperation of 51% is not impossible (Aponte-Novoa et al. 2021). At one point, one mining pool, GHash.io, built up a majority of Bitcoin's computational power. Although it had no intention of executing an attack, it was a warning (Hajdarbegovic 2014). GHash voluntarily reduced its computations, concerned miners left the pool, and ultimately its governing company quit the mining business and concentrated on exchanging bitcoins for dollars.

Mining pools are not created to form 51% attacks. Hardy and individualistic miners collaborate to pool their computing power and share the rewards among sometimes several thousands of colleagues. Their payouts are correspondingly smaller, of course, but the income is steadier and more predictable (Antonopoulos 2015, 207–210). This seems to be a goal worth trusting for, for a mining pool is built on trust – *despite the fact that the blockchain philosophy of trustless trust says that competitors can never trust* (the second tenet of trustless trust from Chapter 1).

Trusted third parties and moments of centralisation, all contradicting the trustless trust philosophy, abound in the real world of blockchain, as opposed to the hype. In a managed mining pool, an organiser takes a percentage of the gains and is trusted to distribute the rest across the pool. The pool also needs to trust its members, because they can withhold their gains from their colleagues (Velner et al. 2017). If the miners don't want to trust a pool manager, they can set up self-governed pools on a separate, parallel blockchain, called a *share chain*. So the miners not only have to mine the blocks for the main blockchain, but also the shares in the share chain. The two largest self-governing pools collapsed because of poor design, unsurprising given the circles they had to square to survive (Troutman & Laszka 2021).

Thus trustless trust spreads its demands through a weird system that, far from being trustless, has an endless thirst for trust of the abstract systems it spawns.

Exchanges

A blockchain defines a cryptocurrency or token simply by creating a database of its transactions. Its value is purely what people are prepared to accept in exchange for it. Many cryptocurrencies are seen as *bona fide* investments, their high value reflecting high risk. The level of trust required for such an investment is correspondingly high; the investor has to trust that the chain will exist until he cashes out, and that some people will always be prepared to receive the cryptocurrency in payment.

To get one's money out of the blockchain and into the real world, one needs an exchange which takes dollars and converts them to the cryptocurrency, or *vice versa*. Handing over dollars involves trusting that the exchange will indeed make the conversion, that the cryptocurrency is genuine, and most importantly that dollars will be returned when the investor exits. Trustless trust ensures all investors can access an encrypted copy of the blockchain, but inspecting it demands resources (that an individual is unlikely to possess), and even so checking the chain doesn't ensure that it is genuine. Yet again, trustless trust needs a good dose of ordinary trust to function.

Exchanges are an undoubted weak point. There have been a number of frauds, as well as genuine exchanges with poor security that were hacked and their cryptocurrencies siphoned away. At Christmas 2021, a laudatory article profiled "the four most powerful people in crypto," "crypto demigods" blessed with "Brain power, idealism splashed with opportunism and physical stamina." Of these four, Sam Bankman-Fried, founder of FTX, "a cryptocurrency exchange … where high finance meets teenage chaos" was convicted in 2023 of wire fraud, commodities fraud, securities fraud, money laundering, and campaign finance law violations; Changpeng Zhao, CEO of Binance, the world's then-largest cryptocurrency exchange, pleaded guilty to money laundering charges in 2023; and Arthur Hayes, former CEO of BitMEX, which specialised in cryptocurrency futures, pleaded guilty to violating the Bank Secrecy Act in 2022. Only Brian Armstrong remains standing. Let me suggest, without proof but with common sense, that it is no coincidence that Armstrong, "the introverted, unflappable CEO of Coinbase, the only exchange listed and regulated in America, is playing the long game, hoping that being the most above-board will help lure punters wary of trading through less-policed offshore exchanges" (Economist 2021). Er, yes. You have to trust something, or someone.

Hence as more investors and criminals have piled in simultaneously, regulators have moved away from a *caveat emptor* approach to regulating exchanges as types of bank (van der Linden & Shirazi 2023). This has helped stabilise the field, to its (and its investors') benefit – but at the cost (to blockchain mythology) of replacing trustless trust with its old enemy, the *trusted third party*, the entity that blockchain was created to make redundant.

Smart Contracts

Contracts were briefly introduced in Chapter 2. They are agreements between at least two parties that set out rights (e.g. to a payment) and duties (e.g. to supply goods or services of a specified standard by a specified time) that the parties take on voluntarily, and agree to make legally binding. Dating back to antiquity, contracts are one of the panoply of means by which we globalise trust in a complex, commercial world with the rule of law. Parties behave in a trustworthy manner by following the contract, and trust because they have remedies to hand in the event of a betrayal.

Odd therefore to position a technological form of contract as *reducing* the need for trust, but in the 1990s it was argued that e-commerce had created a need for contracts to be "smarter" and self-executing (Szabo 1997). A smart contract is basically a computer program triggered when some condition is reached, that runs independently of the parties to it, transferring ownership of assets on a ledger automatically and unstoppably. In other words, it is a "contract" which the parties cannot renege upon, another type of trustless trust.

It is not surprising therefore that the invention of blockchain enabled smart contracts to be implemented practically. The blockchain provided a fixed medium for the program that could not be altered, so its execution would be out of the parties' hands. The Ethereum blockchain[4] became the leader in smart contracts by including a programming language within it that could be used to write smart contracts. The program *is* the contract, and because it can't be altered, breached or broken, trust is unnecessary.

This is all good technohype, but the idea of a *smart* contract rests on a fundamental misunderstanding of what contracts are there to do (O'Hara 2017b). They are *not* mechanisms to make things happen. They are social arrangements, not mechanisms; they share some aspects with promises (Fried 2015). Because they increase confidence in cooperation, they help spread habitual trust and trustworthiness across the economy, in global, not local, form (Kimel 2014).

---

4 https://www.ethereum.org/.

Contracts can of course be broken, but remedies are not punitive, because "a central moral concern in contract law in action is flexibility to recognize the need for adjustment, release, and forgiveness among good faith parties" (Braucher 2012, 667). Parties to a broken contract can demand only the expected direct loss from the party in breach. This doctrine of *efficient breach* (Posner 2014) leads to social benefits, as parties can use their resources to better effect, even after ensuring that the others are no worse off than they were. The law would be paternalistic if it went any further. Hence, contract goes beyond a promise, because promisors and promisees have no legal backing for their agreement, and a promisor is not responsible for correcting the harms she has caused (Shiffrin 2007; Barnett 2012). The courts can adjudicate contracts where necessary, and also strike down unfairness, such as usurious loan rates or an unjustly one-sided employment arrangement (Shiffrin 2000). Fairness is a social construct, and over time the autonomy implicit in private contracts has been reduced by public restraints to reduce injustices, but such changes were never intended to replace or remove trust between contracted parties (Seligman 1997, 100).

Contracts are flexible, and need interpretation, just like the commitments of a trustee. Courts take on the role of arbitrating between parties who fail to agree on their meaning. Flexibility is essential, especially when immensely complex contracts govern large-scale partnerships or projects. But, more to the point, things change and all parties may want and expect contractual terms to evolve over time. The problem with a smart contract – *qua* contract – is its USP, its inflexibility, which will operate even against the present interests of all the parties, having been agreed in the past in possibly very different circumstances.

Smart contracts have their uses, such as allocating resources within a complex organisation or partnership (e.g. the TradeLens platform for managing global supply chains developed by the logistics company A.P. Moller-Maersk, which had great promise, but even that was discontinued as commercially unviable – Jovanovic et al. 2022; Maersk 2022). They can be valuable where trust is low. But *qua* contracts they are dumb, not smart, going *against* the grain of mutable human affairs.

And all this presumes that the smart contract will work according to plan anyway. One particularly interesting disaster happened with an investment company, the DAO (Decentralised Autonomous Organisation), that was made up entirely of smart contracts written on the Ethereum blockchain. It had no management at all (in theory), and investments would be made when a majority of investors voted in favour. The DAO raised $150m in 2016 after much hype (Cleary 2016; Green 2016), at which point crooks exploited a flaw in the software that enabled them to snaffle the funds, now converted into cryptocurrency.

The sequel was interesting (O'Hara 2017b): Ethereum and DAO investors launched a 51% attack *on their own company*, backpedalling down the chain to reach the theft, and then rewriting history to pretend it had never happened. As the crooks had had to park the cryptocurrency before they could convert it back to dollars, the situation was saved.

However, was this ethical or principled? In a smart contract, the code *is* the contract, we are told. Why? So that the contract cannot be rescinded, so trust is not needed. Yet in the face of a hack *which used the code/contract as written*, the smart contract *was* rewritten. But if the code is the contract, *whatever the hackers did was permitted within it.* All code has bugs, so the fact that this code had not worked as its writers intended is neither here nor there. *The DAO* broke the contract, not the hacker. So much for trustless trust.

CONCLUSION: YOU HAVE TO TRUST SOMETHING

Technology can't banish trust by disrupting its forms of life and replacing its practices with mechanisms. "Trustless trust" is an oxymoron. The reason is quite simple. If one replaces a trusted third party with a technical system, then *one must trust the system*: that it has been designed, built and tested by competent engineers, that it is not run by criminals, and that it is secure. At most, all one can do is replace trust-in-trustworthiness with trust-in-reliability. Indeed, crime is the sort of low-trust environment where trustless trust and blockchain may flourish (Juels et al. 2016), but trust is still required for any cooperative criminal enterprise. In 2018, the FBI set up an encrypted messaging service, Anom, that supposedly allowed anonymous and untraceable communication. In fact, all the million messages per day went straight to the FBI, ultimately leading to over a thousand arrests. The crooks, unwilling to trust each other directly, trusted the app (Cox 2024).

Blockchains won't disappear. They have a guaranteed existence as long as there is a network prepared to host the chain, and for cryptocurrencies, as long as people are prepared to exchange the tokens for goods, services or other currencies. Cryptocurrencies will never, in my view, be stable enough to serve as useful media of exchange, and the number of users will fluctuate, up when value rises, down when another scandal happens (O'Hara 2015).

Losses from crypto fraud did decline substantially from 2022, but still amount to tens of billions of dollars annually (TRM Labs 2023; Chainalysis Team 2024). Does that mean cryptocurrencies are getting safer? Yes – but only because they are being regulated by trusted third parties! It turns out that blockchain, which is meant to get rid of the need for trust, *depends* on trustworthy regulation.

We have also seen the importance of information intermediaries. Can, or should, the 20th century ecosystem in which they flourished be restored?

Generative AI is going to industrialise the production of misinformation and deepfakes, but the most potent images from the elections in 2024, for instance, were undoubtedly genuine – Rishi Sunak announcing Britain's election in a drenched suit in the pouring rain, Joe Biden's vacant expression in debate with Donald Trump, and Trump's extraordinary response to his attempted assassination. Much misinformation is spread by fairly visible actors – one could hardly accuse Trump of sneakily hiding behind the scenes (Pfiffner 2020), and neither was Boris Johnson's "chronic aversion to the truth" (Seldon & Newell 2023, 567) disguised very heavily. Misinformation is also welcomed by those who receive it (McGranahan 2017). And in any case, a lot of election interference takes the form, not of misinformation, but of channelling the most polarising (genuine, true) content to those most likely to take partisan positions. Opinion is not often changed by the news we receive; rather, those with the strongest views tend to seek confirmation, and become even angrier.

Resistance to disruption of future elections depends on learning from over-willingness to trust untrustworthy or bad faith news sources. We shouldn't assume background conditions can't change. Those who took to the hills because they were fooled by Orson Welles' *War of the Worlds* radio show in 1938 probably weren't too terrified by *The Blair Witch Project* in 1999. A more sensible deployment of trust, scepticism and mistrust is likely to save us from the deepfakes, now we are aware of the danger.

Trust is likely to save us; technology won't. *Trust is not the vulnerability* – tech is. That is not to say that technology is inherently misleading, or that it won't make our lives very much better. Rather, the real vulnerability is its disruption of the legibility and contextual integrity of existing forms of life. As we deploy technology, we need to compensate for their degradation.

10. Democratic liberalism against trust

This chapter will consider ways in which the valorisation of individual autonomy in democratic liberalism handicaps trust in modern societies. We shall begin by looking at the "official" story of liberal democracy as the institutionalisation of rational mistrust of self-interested political actors. Next, however, we consider how liberalism is over-trusting of the capacities and inclinations of individuals, and compensates for this with covert social coercion. Whereas conservatism combines cautious trust in authority with awareness of human imperfection, these two sections describe liberalism's *mistrust* of authority and *ideal* of human perfectibility.

We then examine the social underpinnings of liberal individualism, democracy and capitalism, spelling out oft-cited criticisms that liberal policy tends to undermine its own foundations. With respect to our own concern with trust, liberalism's deleterious effects on social legibility are the most germane, and are then examined in more detail. As a coda, we consider the recent philosophy of post-liberalism, arguing that it is not what is required in Western democracies, throwing away as it does the civil baby with the antisocial bathwater.

LIBERAL MISTRUST OF POLITICS

Liberal democracy puts the self-interested individual into a self-governing environment. 19th century opponents worried that this arrangement would lead to anarchy and the breakdown of civil society: that clearly did not happen. The benefits have been many, prosperity has been large and widespread, and society has improved on many measures (McCloskey 2010).

In which case, what ambitions could democratic liberals possibly still have? Sofia Näsström undertook an analysis of democracy in the style of Montesquieu (1989), to determine its "spirit" (Näsström 2021). It began from the observation that democracy cannot justify itself, because the composition of "the people" who have the power to decide cannot itself be decided democratically, since there is no prior body to make that decision. For the conservative, this is neither here nor there – if "the people" have a good enough pedigree of making decisions and inhabiting democratic institutions, the system has proved itself, and we are not creating a society from scratch. For the liberal, however, the inability to demonstrate legitimacy is a problem.

Näsström posited *emancipation* as democracy's spirit, a perpetually unfinished project of increasing inclusion in "the people." Different groups get emancipated through time, and so democracy becomes increasingly open and equitable. Authorities that seek to exclude eventually become targets for democracy, and (to the extent that democratic principles triumph) one by one they are eliminated. This creates perpetual uncertainty which needs to be renegotiated by each political generation anew. No-one, and no group, decides who is in and who is out, and stability is never reached. She argued that democracy manages itself through intermediary bodies that prevent any groups becoming dominant (Näsström 2021, 173–174).

From Liberalism to Neoliberalism

Now, however, a tension emerges. As more people are emancipated from ascriptive roles (e.g. as carers, or unpaid workers within the household), either the private sector must take on these roles in a market for services (creating inequalities), or the state must take them on. If more people are to be emancipated from risks (e.g. unemployment, illness), these too must be socialised, either in the private sector via insurance or cooperatives, or by the state. If the private sector is mobilised, the result is atomisation and privatised responsibility, often associated with neoliberalism or what Näsström called the choice revolution (2021, 158–193). The state's intervention creates the welfare state (Timmins 2001), which leads to surveillance of those receiving services (to ensure that public money is properly spent – Gilliom 2001; Coles-Kemp et al. 2018), the shrinking of the private sphere (because public services tend to crowd out the private sector), and an increasing burden on the public finances.

What is often called "neoliberalism" is really the unpleasant discovery that the economy doesn't generate enough surplus to fund the perpetual expansion of a bureaucratic state. Governments are forced to slow the growth of the state (rarely reducing its size) for reasons of fiscal probity, so that more of the social welfare burden falls on the private sector. The costs of emancipation are then paid by the emancipated, self-interest takes over from socialisation, and the individual begins to look like a rather unpleasant fellow, whose "impatience with regulations and red tape and with the laws of the state … fosters an almost perfect identification with a model of society cast as a vast market in which anything and everything can feasibly be an object of exchange, where money, sex, and power are at one and the same time goals and prizes of private and public success, and where the language of the public good seems itself to have gone missing" (Urbinati 2015, 7). Hence, the choice of a welfare state over neoliberalism only postpones the imposition of neoliberal policy.

If there are marginalised groups, emancipation is the job of everyone. Yet citizens also feel the pull of their own self-interest and that of their in-groups, and they may see their main duty as protecting their positions from dilution by newly-emancipated groups. Liberal individualism now looks less emancipatory than egotistic.

We end up with two kinds of liberalism. O'Sullivan distinguished between the relatively relaxed liberal democracy of the 19th and 20th centuries, which aimed to guarantee constitutional protection for liberties such as free speech, association, markets and media, while accepting self-interest, and hyphenated liberal-democracy, which insists on the emancipatory programme (O'Sullivan 2016). The dialectic between these two shapes the operation of liberalism in the real world.

Democracy and Mistrust

Begin with the individual egoist who "doesn't give a damn" (Urbinati 2015, 1–2). Entrusting a specialist core of Burkean representatives with power is not a good idea if they may be such people. On this brand of liberalism, believing that someone will work for the public good rather than her own self-interest is impossibly naïve, a view satirised by Peter Hennessy as the "good chap" theory of government, "that those who rise to high office will be 'good chaps', knowing what the unwritten rules are and wanting to adhere to them, even if doing so might frustrate the attainment of their policy objectives, party political goals, or personal ambitions" (Blick & Hennessy 2019, 5).

"Good chap" is a beautiful name for the theory, evoking a P.G. Wodehouse world run by silly asses with names like Bertie Wooster and Catsmeat Potter-Purbright. The antidote is to limit the discretion of people in public life with a written constitution. "It may be a source of regret for some, but certain elements of the venerable perhaps romantic 'good chap' state of mind need now to be codified in cold hard prose" (Blick & Hennessy 2019, 4). The US might be a model: "the makers of the US Constitution did a fairly good job of designing a system to replace colonial domination and misrule" (Baier 1994g, 161).

The overwhelming consensus is that democracy "presumes distrust of authority" (Uslaner 2002, 222). Who guards the guardians? – "the people must and can" (Hart 1978, 208). We must "cease to put our trust in princes, and instead take charge of the public arena ourselves" (Skinner 1995, 237). "Distrust may be the major engine for an even more democratic state" (Levi 1998, 96) because it "sustains the sense of monitoring and critique, a veritable tonic for the institutions" (Urbinati 2015, 181). "Where such positions of command are deemed unavoidable, then vigilance and nonparalyzing distrust will displace judicious trust as the functional virtues of that sort of highly asymmetrical cooperative scheme. Even when we agree that we need them, we should maintain some continuing distrust of those institutions that create commanders; if we have learned anything from our individual and collective lessons in regrettable trusting, it is surely that power corrupts" (Baier 1994h, 193).

Yet reducing politicians' discretion and monitoring them within a legal framework can't solve the problem, because, *pace* Hennessy, "good chap" naïvety applies to written as well as unwritten constitutions. The United States' constitution, over 200 years old, was easily ripped up by President Trump, the Republican Party, and compliant courts; prosecuting him for his alleged crimes only strengthened him. If "the ongoing practice of effective democratic politics relies on … 'choiceless' trust as a matter of course" (Lenard 2012, 43), then we are in trouble, because choicelessness is not determined by the constitution and its "choice architecture." Choiceless trust is far weaker than the blockchain-inspired idea of trustless trust, because "choicelessness" is in the mind of the beholder. Only if the politician *feels* constrained by the constitution (is a good chap) can she *be* constrained by it.

If you think that a written constitution will moderate the behaviour of an Orbán, a Netanyahu or a Modi, then you are as naïve as someone who thought that an unwritten set of conventions might have constrained a Johnson. If the good chap theory is to be banished, we need to junk the notion that political actors can be restrained at all in a democracy, and we are back with the agonism of Carl Schmitt and his acute diagnosis of the weakness of constitutionalism – a cure somewhat worse than the disease.

If written constitutions can't eliminate good chap naïvety, then how safe is it to place trust in politicians? Lenard argued that trust is important for democracy to function, so distrust of politicians is correspondingly problematic (2012, 11). Hence her distinction between mistrust and distrust. Mistrust is preferable, the scepticism of an active citizenry. "Mistrusters experience an unwillingness or inability to commit to trust or distrust, and … are therefore highly susceptible to new information" (2012, 58–59). Mistrusters don't jump to conclusions, approaching representatives with caution and vigilance, anticipating disillusion (Hirschman 1982, 8). Pierre Rosanvallon discussed *counter-democracy*, where institutional politics is kept honest by frequent interventions from extra-constitutional sources, such as demonstrations and hacktivism (Rosanvallon 2008), while Nadia Urbinati's *counterpolitics* turned disillusion into an active force expressed in dissent (2015, 183). As noted in Chapter 8, Steven Gerencser suggested it might be a means of tidying up Oakeshott's theory of the state (Gerencser 2000, 153–165).

If vigilance takes the form of dissent, won't it simply descend into oppositional beggar-thy-neighbour politics? Realism about the corruptibility of the political process is sensible, but it is paradoxical if not outright contradictory to entrust politicians with legislative powers if we don't trust them (Lenard 2012, 67–68). Lenard's suggestion was that voters' vigilance be founded on generalised trust of each other (2012, 61–62, 71–72), citizens constantly monitoring representatives to ensure that everybody gets their rights. This is consistent with Näsström's expansive emancipatory model, but is surely idealistic. More realistic is monitoring representatives to ensure, for example, that they are not giving rich people tax exemptions, or poor people too much welfare, or immigrants the full panoply of public services, or creaming off public money for themselves. That, however, would hardly meet the emancipatory *desideratum*, while such targeted vigilance need not rest on generalised trust (in fact, seems to require the opposite).

These problems are inherent – we either have good faith and work within the written or unwritten rules, or we are back to Schmitt and his friends and enemies. Counter-democracy, valorising dissent and resistance, has unfortunately but demonstrably nurtured the Peasants. Liberal theorists approve of protest from the radical left, but seem surprised that the more numerous radical right has learned the same techniques. As so often, the Puritans enable the Peasants.

Can the Private Sphere Police the Public?

However the balance is struck, on the liberal blueprint mutual trust among citizens is to be augmented by constitutional arrangements promoting vigilance and dissent to discourage the corruption that essentially accompanies power. The problem is that such connections are bi-directional, and we may have the perverse result that instead of the private sphere policing the public, the public sphere ends up dominating the private instead.

The American founding founders assumed that public opinion would find its way into political thinking thanks to a widespread common culture, but the liberal aspiration of a plural society of autonomous individuals, combined with the democratic spirit of ever-growing emancipation creates greater diversity among "the people." The liberal ideal means that a common culture must be transcended, not relied upon. Worse, to the extent that there is a common culture, it will restrain the autonomy that is the liberal ideal.

Hence, generalised, indiscriminate trust between citizens becomes the means to subvert common culture; "it is not only the direct practice of political power that denotes democratic citizenship but also our *knowing how* we may practice politics, and our *feeling ourselves* to be equals in sovereignty, whether we participate in it directly or not" (Urbinati 2015, 176, her emphasis). Burke's little platoons, and de Tocqueville's associations of civil society cannot be the bridge between the private and public spheres, because they are "often far from respectful of the autonomy of judgment of [their] members … [R]ather the political system itself, with its representative institutions, rights, and impersonal norms … educates and inspires a dignified moral life in all circumstances involving action, even in ordinary and private ones" (Urbinati

2015, 174). While liberals valorise autonomy in the public sphere, they also insist hegemonically that the private sphere respects it too.

Yet this arrangement simply means that the impetus for vigilance comes from the public sphere, through the political system, and private space correspondingly shrinks (O'Hara 2023, 193–197). Not only must politics be inclusive, but non-political groups – families, churches, schools, universities, football teams, companies – must be too (Legutko 2016, 20–21). John Stuart Mill was concerned about informal social pressures on individuals, but far stronger are the pressures from liberals promoting autonomy. Beate Rössler, for example, highlighted privacy's role in making personal autonomy possible, but was also concerned about the denial of women's autonomy within the household, and so rejected arguments that private space should be free from public scrutiny (2005, 106–110). Everything legitimately done within the home should be protected (2005, 124), but as most norms of private life reproduce the unequal gender order (2005, 158), not very much is legitimate.

Rössler was particularly scathing about the analyses of Locke, Mill and Rawls, who saw the household as a black box whose structures and inequalities should be unquestioned (2005, 27–33). But the household is not the only target. A private gentleman's club, the Garrick, was browbeaten in 2024 into accepting women; no sooner had the freedom of association of the Garrick's members been crushed, than commentators were already complaining that the new unisex association "remains an elite club where only a chosen few are invited, and even fewer can afford membership" (Olchawski 2024). Excluding anyone at all is sinful in this hegemonic liberalism, even from a private association. Women's spaces have been similarly pressured (Brunskell-Evans 2020, 93–97; Stock 2021, 5–6). Liberal commentators constantly criticise religious institutions that affirm traditional doctrines of gender, sexuality or abortion, even though, on classic liberal doctrine, it is none of their business (Legutko 2016, 172–175). Hence the liberal project of creating inclusive political space morphs into the illiberal one of insisting that institutions, traditions and practices are inclusive – but since smaller associations must be exclusive to survive, this is tantamount to their destruction.

The Burkean model of representative democracy assumes that citizens entrust the political work to a small service cadre of representatives, leaving them free to concentrate their energies on their private business. But under emancipatory liberalism and counter-democracy, the private sphere is supposed to feed back into the political world via citizens' vigilance, dissent and extra-constitutional activism. Hence, the specialist representatives are obliged to expand their activities to police the private sphere itself, to ensure that liberal principles hold there as well, becoming the organising principle of the whole of society, not only the political sliver.

LIBERAL TRUST OF THE INDIVIDUAL

Why does this happen? Excessive mistrust characterises liberalism's view of the democratic process, but excessive *trust* pervades its model of the democratic *citizen*, constantly engaged and vigilant against the centres of power. Its idealisation turns sour fairly quickly. "Liberals cheered the degradation of our popular culture, and then professed to be surprised when it spilled over into our politics. They shouldn't be. If they're upset, they're experiencing the rage of Caliban at seeing his face in the mirror" (Buckley 2018, 110). In their dismay at how individuals exploit their autonomy, their instinct is to nudge, steer and control (Crick 1964, 77; Legutko 2016; Pabst 2021).

Liberalism abstracts individuals from their relational essence, concealing what makes them unique behind a "veil of ignorance," and downplaying their histories, relationships, dependencies, ideologies, nationalities and ethnocultural inheritances. Hence, if a social structure is to be a legitimate preference of such an etiolated individual, it must provide benefits not only to that individual, but to others equally; only in these Kantian terms can autonomous abstract individuals justify such structures rationally. The individual's vigilance about democracy therefore should take the form of ensuring that no-one is treated differently. Because their view of an individual is so abstract, "many liberals fall into the trap of believing that the historical, cultural and institutional context of government is irrelevant. Institutions and traditions that impose obligations on us can simply be cast off" (Timothy 2020, 35). Liberals expect people to assert their self-interest by searching for the common good, without appreciating that group interests are the corporate means by which individual interests arise and are pursued (Crick 1964, 127).

As feminists have pointed out, an autonomous individual most resembles the men of business who populate social life as equals: "self-sufficient, independent, and self-reliant, a self-realizing individual who directs his efforts towards maximizing his personal gains. ... Talk of rights, rational self-interest, expediency, and efficiency permeates his moral, social, and political discourse" (Code 1987, 358). Independence and the universalisations of the categorical imperative outweigh particularised trust, loyalty, friendship and caring, while rights are emphasised over responsibilities and duties. This is simply unrealistic as a representative view of humanity (Baier 1994e, 114).

Kant's call "*Sapere aude*! Dare to know!" was a rallying-cry for the self-reliant individual, one of the slogans of the Enlightenment. Take nothing on trust, take courage to find things out for yourself, liberate yourself from your immaturity. Repudiation of the spurious heritage of convention, tradition and superstition has been called "a constitutive virtue" of democracy: "Dissent and sovereignty of judgment are not simplistically cast in an antiauthority role or as a reaction to the power of the majority but also have a role in consolidating self-culture; they are therefore at once a private and a public virtue" (Urbinati 2015, 185).

Unfortunately, even virtuous dissent can be wielded against the epistemic authority of rationality, reason and scientific method. Sociological critiques of science provided the seed for those who dare to know to stop deferring to reason. Instead, one may adopt a self-conscious open-mindedness that refuses to be "closed down" by scientific orthodoxy (Hobson-West 2003; Baker 2022), or explore alternative and more congenial sources of authority, such as feminist or neopagan ideas about the protective, intuitive or doting mother-figure (Baker & Walsh 2022). Such people tend to label themselves "critical free-thinkers" (Harambam & Aupers 2017). The most conspiratorial eschew epistemic authority altogether, so that "daring to know" presents itself as being able to face down an infinite regress of suspicion (Knight 2000).

The point is not that liberalism entails such stances, but rather that it has surrendered the tools for reasserting epistemic authority against them. In such an environment, it is a stretch to argue that "Even though they corrode the social sentiments, [dissent and sovereignty of judgment] reinforce sympathy and cooperation between citizens since … we discuss and wax passionate about the things which we love and to which we are bound by ties deeper than rational assent" (Urbinati 2015, 185). Certainly there are deeper ties than rational assent, but do they necessarily preclude respecting epistemic authority? Urbinati seemed to equate such respect with conformism: dissent and sovereign judgment "mitigate the tendency to cultural uniformity that is inherent in democratic society and bolster acceptance of majority rule as a method of making decisions that is based upon a recognition of the equal fallibility of citizens, and of their right to review opinions and decisions taken" (Urbinati, 2015, 185). But the inherent tendency to cultural uniformity has functional value, underpinning trust, and therefore sympathy and cooperation between citizens.

Relying on our own sovereign judgment is just as likely to turn us against as toward our fellow-citizens, seeking a shared antagonistic social identity with fellow conspiracists (Mikhaeil & Baskerville 2023), and in the worst, though rare, cases, perpetrating extremist violence (Pierre 2024). One unrepresentative study of college students suggested a link between conspiracism and low self-control (Heitkamp & Mowen 2023); those who struggle with self-control tend to be more impulsive, insensitive to others, welcoming of risks and criminally inclined, and are therefore precisely the sorts of people who benefit from exposure to conformist pressures and uniform cultures. By all means celebrate difference, but not all difference is good or virtuous.

Put another way, there is a concealed asymmetry. Urbinati warned against a "haughty tolerance of the mad," and recommended "equality as a value, the feeling of sympathy, curiosity … toward what is different – perhaps to criticize it, or to welcome it, and often to imitate it – … [impelling one] in the direction of dialogue and a readiness to listen" (Urbinati 2015, 157). All dandy, but liberals want people not only to make choices, but also to choose well (i.e. what liberals want them to choose). Trust enters the liberal story here: voluntary compliance with democratic norms depends on trust that others will also comply (Lenard 2012, 53), but in trust's absence democratic institutions will not be able to secure the compliance they need (Lenard 2012, 42).

This presents a problem. Individuals are expected to use their autonomy well, choosing good laws, complying with beneficial norms, avoiding free riding and looking after groups they are not part of (Lenard 2012, 39). But if the individual asserts her sovereign judgment and dares to choose something else, the private sphere fails to reflect the ideals of the public sphere, and liberals warn of nationalism, intolerance or xenophobia, "even if these are only suspected or felt rather than actually perceived" (Legutko 2016, 5). The problem then is how to engineer that people make the right choices as defined by the liberal without breaching their autonomy.

The problem is created by liberalism's focus on choice. If we understood virtue in terms of habits and norms, and social behaviour in terms of existing forms of life and the practices of trusting and being trustworthy, it is clearer how compliance with norms often comes "for free," apparently chosen without demur. Of course, "rightness" would then not be defined by the liberal, but would be immanent in the community's practices.

For instance, Hollis wanted to "realise the power of the social sciences to influence [people's understanding of their moral and social world] by putting ideas into people's minds" (1998, 156–157), in other words creating a paternalist discourse to seed minds through education and the media with externally-derived ideas of goodness and virtue. This covert approach to "improving" everyone was prompted, in Hollis' case, by his discovery that trust and trustworthiness cannot be based entirely on rational premises. This is true, but the question he failed to answer is why, given trust and trustworthiness are so widespread, we have to brainwash people into trusting and being trustworthy at all? The motive, of course, is the liberal's basic aim of neutralising traditional sources of authority and undoing convention.

Hence liberalism's bias toward generalised trust, disregarding trustees' personal characteristics. It suggests, among other things, that freedom of association must be curbed, because it promotes particularised trust between people who have common cause, and exclusion of those who don't. If it is forbidden to exclude people, one's associations become weaker, as one must be prepared to trust all comers. In that case, one will be less inclined to take risks and place trust in the association or its members.

If trust is taken in this inclusive sense, then it becomes the vulnerability many make it out to be. Particularised trust is far more secure, because it can be graded depending on one's informed view of the would-be trustee. If one is forced to take people as the idealised trustworthy individuals of liberalism, one is obliged to make assessments based on less evidence, avoiding hard discussions and challenging negotiations. Yet real-world people are not perfect, and we should be allowed to face that. "Politics involves genuine relationships with people who are genuinely other people, not tasks set for our redemption or objects for our philanthropy. They may be genuinely repulsive to us, but if we have to depend upon them, then we have to learn to live with them" (Crick 1964, 124).

THE SOCIAL UNDERPINNING

This counterproductive mix of disappointed trust in individuals and cynical mistrust in public figures risks undermining the social arrangements, customs, traditions, institutions, practices and the rule of law upon which liberalism relies, hollowing it out (Kekes 1997; Garnett 2004). Whereas Burkean conservatism both depends on trust and is intentionally supportive of it, liberalism rests upon trust and social legibility, while eroding both.

Much depends on the dogmatism of the liberalism on show. A mild version that focuses on preserving existing liberties within a well-understood democratic and market-oriented social framework ("what we do around here") will be more flexible and sensitive to the needs of trust than more combative versions that double down on abstractions such as individual sovereignty, free market fundamentalism, moral relativism, unrestricted globalisation and identity politics (Hollis 1998, 117–125; Lenard 2012, 48; Urbinati 2015, 7, 175; Timothy 2020, 32–34, 121; Pabst 2021, 69–79).

Liberalism is, in Alexander Rüstow's phrase, sociologically blind (Rüstow 2017, 154), but not thereby sociologically impotent. It pretends to a neutrality that plays no role in shaping character, yet influences temperament by removing irksome constraints on behaviour, and facilitating hedonistic and material pleasures (Deneen 2019, 5). A democratic culture that "protected the pluralism of spheres of life" might indeed be useful for nurturing new forms of belonging (Walzer 1983; Urbinati 2015, 18–19), but if it is a genuine commitment to plurality, it can't therefore nudge people away from more traditional and ancestral "spheres" or forms of life in order that new forms of belonging materialise, without risking the loss of the blessings of the older forms.

Mill originally described a world in which individuals' resistance to their traditional constraints was a perpetual and fundamental social fact: "The struggle between Liberty and Authority is the most conspicuous feature in the portions of history with which we are earliest familiar" (Mill 1991, 5), yet the ultimate victory of the former has been somewhat Pyrrhic. "The affirmation of civil rights … has emancipated individuals from preexisting social, authoritarian, and hierarchical snares but has not consolidated new ties consistent with the culture of rights; it has not produced the sort of ethical cement capable of binding together a society of autonomous individuals. Emancipation from the strictures of an authoritarian morality has produced a practice of liberty as license and a vision of individual liberty as antisocial individualism" (Urbinati 2015, 17). The reciprocity of generalised trustors will not be enough to secure trust unconditionally in a society without authority (Gray 1995, 112–117, but for the opposite view, Seligman 1997, 89–91).

Institutions underwrite much of the stability that enables liberty to flourish in a pro-social way, but of course they do this by curbing or limiting the behaviour of their members and users. To the extent that they are informative, surveys have tended to confirm that levels of trust in institutions correlate positively with values stressing stability, protection, and preservation of traditional practices, and negatively with preferences for radical action and change, and moreover that the religious and conservative express more trust in institutions than the non-religious and left-wingers (Devos et al. 2002).

Sub-surface and often ascriptive social structures, especially the family, are important for socialising people as workers, collaborators and colleagues, raising and educating children, and other vital functions (Deneen 2019, 30; Timothy 2020, 82, 120). Even some prominent feminists accept that various gendered structures play beneficial, and possibly irreplaceable roles, while reasonably demanding that they be properly acknowledged and compensated (Baier 1994a, 6). Yet liberalism deliberately works to undo these structures, because they are often unconsented, unpaid and gendered (Rössler 2005, 27–33 has already been cited). By undermining its bases, liberalism places greater demands on trust while simultaneously subverting it (Seligman 1997, 97), wanting the fruits of politics – liberty, representative and honest government, prosperity and general education – without the gnarled tree (Crick 1964, 123).

When the bases of a liberal society have been undermined, the state must step in to replace what is lost. "Ironically, the more completely the sphere of autonomy is secured, the more comprehensive the state must become" (Deneen 2019, 38).

AUTONOMY VERSUS LEGIBILITY?

For liberals, freedom is necessary to facilitate social experimentation, therefore uncovering new forms of life and ways of behaviour, and to allow dissent, thereby challenging received wisdom and revealing new truths (Gray 2000, 105; Timothy 2020, 54–55). While there is an element of plausibility in this, it is an asymmetric vision. The trial and error of the past has produced customs, practices and institutions that are now familiar, legible and functional, yet the liberal rejects these, enamoured only with the potential of the future.

To insist that only future speculations are valid, while past discoveries are illegitimate types of authority, seems unmotivated and unwise (Deneen 2019, 146). Mill's phrase "experiments in living" dignified hedonism with the prestige of science (Kimball 2005, 223), yet, unlike real experiments, they aren't confined procedures to validate hypotheses. They have effects and consequences that ripple through social networks, which cannot be controlled and need not be positive. In this section, I will review the relation between conformism, legibility and trust, before arguing that liberalism in any of its forms depends crucially on legibility.

Conformism

Conformity is a way to connect with people, and nonconformity puts up barriers. Institutions that apply normative standards of dress, manners, etc., give us formalisms for signalling our trustworthiness. Liberals warn against "confusing a shared acceptance of the democratic foundations with conformism and an unthinking adherence to the most widespread or general opinion" (Urbinati 2015, 172), but there is no reason why conformity should negate critical thought (Schoeman 1992, 5–6; Kekes 1998, 128–131, 149–150). Equally, parading one's rejection of social norms as a badge of honour isn't very bright if one wants to make friends and influence people.

For instance, the BBC ran a news item about a man who had a prominent facial tattoo, which had altered his life in various ways, not least his relations with his mother and his employment prospects (BBC News 2015). His girlfriend was shocked, but got over it. Only his daughter garnered liberal praise in the story for being non-judgmental. The BBC view was that all stigma is wrong, and that the person beneath the tattoo may have impeccable character, and is only expressing himself.

The person beneath the tattoo may indeed be a fine fellow, but is his own worst enemy. In the first place, it is hardly a secret that facial tattoos are often perceived as scary, and make people nervous, not receptive. They also clearly proclaim the wearer's deviation from the common run. What's wrong with that? Nothing, except folk who make a point of deviating from social norms render their behaviour less predictable and their attitudes and motives obscure. While they egotistically celebrate their uniqueness and difference, it is harder for them to signal their trustworthiness, be they ever so trustworthy, to the majority. And unlike, say, a Mohican haircut, a tattoo is a permanent obstacle.

From which it follows that it cannot be the case that someone with a facial tattoo, in a society where they bear a stigma, has no character flaws. He has at least one: he lacks prudence. That in and of itself would rule him out of many careers – who would want an imprudent bank clerk, or General, or judge, or dog walker, or swimming instructor, or childminder? Of course, getting inked could be helpful in some careers – not only as a tattoo artist, but also, say, a social worker dealing with high-deviance groups. Such a person may be more easily accepted by those with a low level of trust of stuffed shirts and squares. But even there, the appropriateness for the job or otherwise of the tattoo *depends on the clients*, not on the person seeking the role. Roles come with expectations attached; we can of course work to remove them, which is part of the BBC's liberal social engineering agenda, but they can't be wished away.

Luhmann correctly argued that "Whoever presents himself from the outset as unapproachable – … [for instance] by offending against customary demeanour or behaviour in a way which shows that one places no value on it – … is in no position to acquire trust because he offers no opportunities for learning and testing" (2017, 69), but went on directly to say that mere conformism cannot provide trust either, because "role-conformity offers little opportunity for the presentation of oneself" (2017, 69). Luhmann may not have meant this, but it follows that the way to signal trustworthiness within a role is to *outperform* expectations of it, not to repudiate its restrictions entirely. The important thing is not *confounding* potential trustors, but *impressing* them, which involves being other-directed, taking their prejudices, habits, expectations and preferences seriously.

Liberalism promotes the freedom of the trustee, but by doing so frustrates the interests of the trustor, and thereby those of the trustee. The problem is that the trustor is unable to muster sufficient confidence, because he isn't getting enough valuable information from the would-be trustee's signalling. If she is not legible to him, cooperation and collaboration will be hindered.

Legibility

Illegibility has attractions for liberals. For instance, the relative illegibility of close communities to outsiders affords protection from the controlling impulses of bureaucrats from the capital (Scott 1998, 54, 303). Conversely, when we become legible, we are exposed.

Yet legibility is necessary for overarching liberal objectives. In a just society, Rawls argued that everyone would have reasonable confidence in their ability to pursue their projects and achieve their goals, which would be enabled by a more even distribution of resources and capital (1999, 155–157, 386–387). However, in his emphasis on distribution (indeed, overemphasis, as argued convincingly by Kekes 1998, 178–185), he overlooked what would appear to be an absolute precondition for such confidence. As well as providing sufficient resources and liberty, surely the just society's institutions, practices and conventions *should be legible to the individual.* Each person should be sufficiently aware of how their society works, so that, for instance, they have a sense of how to get things done, what others expect of them, what norms obtain, how a contract is likely to be enforced, how patterns of ownership and control play out, and so on. There would be no point in creating the perfectly just institution if it violated contextual integrity, could not be integrated into ordinary life, and was incomprehensible to citizens.

Change and innovation, however well-meant and rationally justified, will impact on legibility, which in turn will impact on Rawls' own project of creating a just society. It follows that, even if a perceived injustice is sufficient to legitimise social engineering and restructuring, there are epistemological reasons that should provoke suspicion of top-down change, particularly from a rationalist state bent on creating "justice" or "efficiency."

One alternative to Rawls' approach has been to focus not only on the resources, but also the *capabilities* required to achieve one's goals. Amartya Sen argued that, whereas Rawls focused on outcomes, the processes and agencies which produce the outcomes are just as important, partly because of varying ideas of what constitutes welfare or "good outcomes" (Sen 2010). Martha Nussbaum took this approach further by enumerating a set of core capabilities, including the ability to engage in critical reflection on the planning of one's life, as well as control over the environment in terms of rights to political participation, free speech and association, and property rights on an equal basis with others (Nussbaum 2011). Can the capabilities-based approach avoid Rawls' legibility problem?

It is certainly more sensitive to context, but it surely requires the same precondition that the institutions, relations, rules and practices of the society be legible to the individual. Legibility means that the individual can craft a plan and a set of goals based around what he is capable of, and work out what his capabilities could actually achieve in the specific context. His control over the environment, be it ever so omnipotent, cannot achieve this on its own, because it will impose a huge cognitive overhead if it is not legible to him. It is one thing to know what he can do, and quite another to know what can be done. The former is his capability, and the latter depends on legibility.

Hence liberalism requires legibility, which is essential both for trust and to enable individuals to pursue their life projects successfully. At bottom, this is the epistemological aspect of the social underpinning of trust that liberalism simultaneously relies on and threatens.

CODA: POST-LIBERALISM AND THE LIBERAL SOCIETY

In recent years, the philosophy of *post-liberalism* has emerged (Gray 1993b; Deneen 2019, 2023; Pabst 2021; Rose 2021, 1–17), with a similar critique to this chapter's of the problems immanent in liberalism's sociological blindness. It is intended to correct liberal error by focusing directly on measures that preserve communities, including somewhat statist economic and industrial policies, the pursuit of a common good defined in terms of local solidarity, the rejection of internationalism, and the redefinition of "liberty" to exclude licence and emphasise self-rule and self-discipline. In the culture wars, it is firmly against the Puritans, "progressives" and the elites, but, though inclining to and sympathising with the Peasants, usually tries to avoid aligning directly with their revolt, while perhaps cheering on as they kick the door in.

Space precludes going deeply into it, but this book is not meant to be a contribution to the post-liberal project. In short, my starting point is that the liberal democracies of Western Europe are the greatest, most successful societies in the world, and indeed through history. They are not perfect, by any manner of means, but how could tens of millions of people live on small tracts of land without friction, conflict and injustice breaking out here and there?

Liberalism is a fine thing – as long as we understand it not as a universal principle that should be applied to everybody, but rather as a tradition that has nurtured and been nurtured by societies in Western Europe, and that has spread unevenly across the globe, particularly to the Anglosphere and, diluted, to a few other places. Not uncoincidentally, these places vie with each other to be major destinations for migrants, artists, investors, thinkers, workers, refugees and dissidents, who want to go from worse places to better. To the extent that other traditions learn from Western liberalism, they will improve. But if, say, Brazil, China, India, Russia or South Africa, never mind Saudi Arabia, Venezuela or Chad, want to pursue other directions, of course they can, and should. Women's rights and gay rights are excellent and important things, but I can't say that the West has the legitimacy to force them upon those who think them wrong or evil. Autocrats such as President Putin gain legitimacy from posing as the guardians of the traditional family against the inroads of Western liberals (Applebaum 2024).

Having said that, we in the West have the right to criticise as much as we like, and to withhold aid and arms if we wish. We may even use economic muscle to force human rights on poorer places, in sub-Saharan Africa for example. Absolutely fine, but let's at least call it by its name – imperialism – and stop telling ourselves that we are better than David Livingstone and engaged in a different project from his.

A Western-centric view? Sure, but then I am from Western Europe. What do you expect? If you come from somewhere else, you no doubt have a different view – I wish you joy of it, but I don't see why it should change my priorities.

How do I square this positive view of liberal *societies* with my dim view of liberal*ism*? My point is that the liberal tradition needs to be understood conservatively. All the social preconditions for liberalism, including trustworthiness, trust, legibility, stability, freedom of speech and association, markets, contracts, and the rule of law, need to be kept in place for *conservative* reasons – because they exist and have value through existing, *pace* Mounk, who argued that "liberalism is a force of progress, not the status quo" (2023, 261). In particular, the Western liberal democracies are largely high-trust places, not because of liberalism, but because of those underpinnings. The justification for liberalism is not any kind of universal principle derived from its own assumptions, or Näsström's emancipatory principle forever searching for new citizens to set free. It is justified by the good it does, where it does it, in the societies in which it does good. It is, *pace* Mounk (2023, 257), a tradition, not a set of principles. In particular, liberalism justified in these terms will not threaten the bases of trust and trustworthiness which have allowed it to flourish in a few fortunate parts of the world.

11. Critical "progressives" against trust

Critical theory is a branch of social thought devoted to revealing, challenging and undermining power. Originally developed by Frankfurt School thinkers such as Theodor Adorno, Max Horkheimer and Herbert Marcuse, its influence endures on the radical left. While it has mitigated the preoccupation with Marxist orthodoxy that used to shackle thinking on the left, and has allowed heterodox ideas to emerge, it rarely shows a constructive side, anchoring a *smorgasbord* of insoluble problems with a deadweight of historical guilt. Self-styled "progressives" who are its intellectual heirs disdain the pragmatics that must inevitably characterise life in crowded democracies, are unimpressed by historically high levels of evenly-spread prosperity, and see no irony in being paid from the public purse to deliver critiques of Western society, free from the fear of arrest and persecution, and in centrally-heated lecture rooms at that. Critical theory is rigorously critical of everything, except itself, to which it is remarkably indulgent.

This chapter is not solely about the direct descendants of the Frankfurt School, but engages with those whose aim is critique, and whose antennae detect injustice wherever they are pointed. Dissent, valorised by liberals, is the *raison d'être* of critical progressives,[1] and their intended legacy is to make resistance endemic. Their failure to transcend an adversarial posture guarantees a position both frivolous and unsympathetic (Kimball 2005, 222). Their work forms the core of the Puritan credo.

It is no surprise that such an ethos is inimical to trust. Luhmann began his opus *Trust* with a pointed jibe at the Frankfurters' "moral intrusion into sociological concepts … . If one remains at the level of critical demolition and surprising presentation of the familiar in unfamiliar ways or of unmasking ideological beliefs through providing causal explanation or evidence of hidden side-effects, then the disadvantages predominate. The identity of terms so employed is misused in order to discredit their invoked meaning horizon" (2017, 3). In other words, the dice are loaded against the *status quo* from the beginning.

[1] I will drop the quote marks for the rest of this chapter, although self-styled "progressives" have little very progressive about them.

At its worst, the critical attitude combines cynical refusal to see anything other than venality in democratic politics, with artless delight in the machinations of insurgents, leading to partisan and Manichaean politics, naïve adulation of murderous incompetents such as Che Guevara (Brincat & Löwy 2023), and cheerleading for regimes such as Cuba's (Honderich 2005, 328n.24) or Venezuela's (Gott 2011). The barely-suppressed joy in some of the early responses to Hamas' ruthless massacres, sexual violence and kidnapping of Israeli citizens in October 2023, even prior to Israel's disproportionate and quite possibly criminal response, beggars belief (Johnson 2023; Kimber 2023).

Even in democracies, such partisanship eschews trust. In an analysis of some recent chicanery in Spain, Castells denied the trustworthiness of any politician, as "political corruption is in the basic DNA of almost any political system" (2019, 18). But what childish glee when his own preferred faction gained advantage in an internecine struggle within the Socialist Party!

> Against all expectations, the evidence [gathered by the team of Castells' man Pedro Sánchez about a party vote] suggested that Díaz was receiving fewer nominations than Sánchez, but the really interesting thing was that though Sánchez's followers detected this unexpected trend, they knew that they were being watched closely, and so chose to leak contradictory information. This reassured the socialist establishment into believing it had overcome the challenge by doubling its rival's number of nominations. … [It] had no plan or strategy for [the eventual] outcome". (Castells 2019, 115)

In other words, the sainted Sánchez simply got his followers to lie about the information they gathered, "leaving Díaz to pack her bags and retreat to her palace in Seville, muttering about revenge" (Castells 2019, 115). Result!

The lesson we seem to learn from this is that politicians are untrustworthy, and therefore bad and corrupt, apart from the ones on the radical left whom Castells likes. These, on Castells' own account, "defined a renewed socialist strategy, opening the doors to a new kind of left-wing politics in Spain" (2019, 115–116), *precisely by being untrustworthy*! Sánchez used his talent for mendacity to great effect in 2023, when, after repeatedly insisting that an amnesty for those who organised an illegal referendum on Catalan independence would be unconstitutional, he then granted one as a *quid pro quo* for support as he cobbled together a Parliamentary majority. The need for trustworthiness melts away in the heat of partisanship, as each side trusts its own unconditionally.

Castells retained a sentimental attachment to trust, arguing that "the most effective form of political attack is to destroy … trust through the moral degradation of someone's character and image as leader" (2019, 20), and adding that "while the effects of scandal politics on specific politicians are indeterminate, the practice does have a devastating second-tier effect; it inspires a feeling of distrust and moral reprobation towards all politicians and politics as a whole, thus contributing to the crisis of legitimacy" (2019, 21). That is true, and Castells, a master of the form, is demonstrably as guilty as the worst Republican attack dog – thank you Manuel. Our current leaders, on his account, are narcissistic snake oil salesmen, cynics disguised as realists who exploit people's weariness with corruption and injustice to subvert democracy (2019, 131–132). Populism is succoured by implacable critics of bourgeois democracy of both left and right (Müller 2017, 50–52); the Puritans, as ever, enable the Peasants.

THERE IS NO GOOD TO BE HAD HERE, BY DEFINITION

Critical progressives struggle or even refuse to see any good in existing communities, whereas conservatives value them and are concerned to protect them. Even liberals appreciate (many of) them, but erroneously think that they will be unaffected by their social engineering. The liberals' problem is complacency, even negligence – a sin of omission. But the progressives' problem is a sin of commission. Their desire to dismantle oppression and domination means that – since these are to be found everywhere, especially in socialist paradises – all communities must be radically reconfigured, whatever the totality of the evidence and the balance of the harms.

Bourdieu's sociological framework, honed in his nostalgic early work on resistance to modernity in Algeria and the Béarn, is based around three elements of *habitus* (the skills and habits that enable us to negotiate the social world), *field* (the setting for agents and their social positions, analogous to what I am calling the form of life), and *capital* (symbolic, cultural and social assets that can be put to productive use). It is an interesting and useful framework, in particular allowing the expression of the importance of legibility. Yet Bourdieu himself rejected this quasi-conservative interpretation, complaining without justification or argument about arbitrary and unequal distributions of capital wherever he found them.

For Bourdieu, capital, despite its immense value in promoting legibility, is to be challenged because it reproduces relations of inequality (Bourdieu 2018), which it certainly does, though this, if an evil at all, is a lesser one. Arbitrariness in particular seemed to upset Bourdieu (1977, 1991), as if it could be engineered out of culture and society. If the world, however contingent, is legible and familiar, people can negotiate it, which surely matters more to them than an unfamiliar egalitarian utopia.

Capital exists only within fields, and reconfiguring the latter to impose equality would therefore mean destroying what capital a society held. If we take the capital metaphor seriously, its destruction cannot result in its fortuitous reappearance in the revamped field. People adapt their plans to the extent of their abilities and chances (Bourdieu 1998, 216), but the legibility they require to assess this would be annihilated by the radical social change advocated by Bourdieu. In short, in utopia no-one would have any social capital at all. Even so, many critical progressives prefer living in chaos to preserving order on the old terms (Castells 2019, 134).

It is difficult to imagine. This was more or less admitted by Castells, who lamented that no new inspirational institutional forms had emerged recently, and accepted the uselessness of the movements he praised, such as Occupy. His only message of hope was the occasional green shoot in the form of "the old utopian vision of self-management," embryonic projects of people "rebuilding the fabric of their lives, bottom up, in the personal and social spheres" (2019, 133–134). It's not obvious who was preventing them from doing this in the first place.

It needn't be like this. Radicals in the 18th century often positioned themselves as "patriots" by analogy to the Romans, free citizens who loved the Republic because of what it was and what it provided. It is a conditional patriotism, but open to the recognition that there are good things to be had (Crick 1964, 76). Today's democracies have flaws, certainly, but where else do such prosperous and tolerant societies exist at scale?

SYMBOLIC VIOLENCE

Perhaps the most egregious mistake of critical theory is its equation of restraint with violent coercion. Certainly, not all violence is physical. The Nazis' deliberate dehumanisation of imprisoned Jews was a central strategy of the Holocaust, and as a victim of the Chinese gulags in the 1950s wrote, "when you're being beaten up, you can turn into yourself and find a corner of your mind in which to fight the pain. But when you're being spiritually tortured by thought reform, there's nowhere you can go. It affects you at the most profound, deepest levels and attacks your very identity" (quoted in Dikötter 2013, 248).

However, critical pessimism turns this on its head, assuming a continuum between the physical restraint and what even a judicious commentator called the "cage of norms" that constrains actions in polite society (Mounk 2022, 114–115). Michel Foucault argued that all power is exercised by dominating agents, that all restraint results from the exercise of power, and that therefore all restraint, including self-restraint, is domination. Furthermore, all power is effectively equivalent to the power of the totalitarian state; any restriction on action is down to the exercise of power; and there is no real difference between those sources of power with authority or consent and those without. Even language is suspect; discourse is oppressive, as it is controlled by the powerful and imposed on the weak. Language, custom, institutions and norms construct reality for those without power (and for everyone else, actually), so they are forced to live in a world not of their own invention. Equal rights won't cut it, because they only apply within that constructed reality which has not been consented to (Scruton 2015, 98–113; Timothy 2020, 40–41).

This exaggerated position becomes of interest to our inquiry when applied to the bourgeoisie. Foucault's obsession with the bourgeoisie was shared with Marx and Engels, whose *Communist Manifesto* charged that class with appalling crimes such as emancipating the serfs, ending feudal and hierarchical relations, increasing trade, commerce, science and navigation, revolutionising industrial production, showing "what man's activity can bring about," accomplishing "wonders far surpassing Egyptian pyramids, Roman aqueducts and Gothic cathedrals," creating a cosmopolitan society, satisfying material wants, expanding people's horizons, lowering prices, urbanising, and altogether creating "more massive and more colossal productive forces than have all preceding generations together" (Marx & Engels 2002, 219–233). Foucault believed "that anything could be deduced from the general phenomenon of the domination of the bourgeois class" (Foucault 1980, 100).

We have already seen how class war against the bourgeoisie failed. The left's shift from class war to culture war was facilitated by the ingenious invention of "symbolic violence" (Bourdieu et al. 1999), which enabled it to soft pedal the inconvenient fact that the nice democratic middle classes rarely lift a finger in anger, and indeed were largely in ignorance that class war was raging around them. Symbolic violence is the splendidly lunatic idea that any social norm is effectively equivalent to a rubber bullet, the logical endpoint of the "cage of norms." All the conventions and distinctions that define civil society are turned upside-down, becoming restrictions to be removed, excised or otherwise neutralised. Language itself, and even thought, are to be policed, and purified of the violence inherent within them. Microaggressions are equivalent to macroaggressions, except that they are easier to "call out," because the microaggressor is a timid fellow quite likely to shrink under rude criticism, whereas the physically violent are, well, physically violent and jolly dangerous. You go and tackle President Putin, while I mortify someone who has just used a banned pronoun. Sticks and stones may break my bones, but names will always dominate me.

The obvious response, that stressing symbolic violence inappropriately trivialises physical violence, is itself an act of symbolic violence, dismissing the reality of symbolic violence (Bourdieu et al. 1999, 4–5). Indeed, in the furthest nonsensical reaches of the theory, symbolic violence is *worse* than the physical kind. Slavoj Žižek stated that "Subjective [i.e. physical] violence is experienced as such against the background of a non-violent zero level. It is seen as a perturbation of the 'normal', peaceful state of things. However, objective [i.e. symbolic] violence is precisely the violence inherent to this 'normal' state of things" (Žižek 2008, 2). Physical violence and the pain that it engenders are devalued as subjective and relative, whereas objective, real violence is to be found in our language and our routine social arrangements and conventions.

In an especially intriguing comment, Žižek likened what he called systemic violence, "the often catastrophic consequences of the smooth functioning of our economic and political systems," to "dark matter" (2008, 2). Note his equation of the tranquil progress of the democratic capitalist world with cataclysm, and his use of the dark matter metaphor, which gestures towards hidden gravitational forces that help social cohesion. As we have seen, that metaphor is regularly used to apply to trust, and, following Nietszche's argument against trustworthiness, it is not a stretch to see trust, and the pressures to be trustworthy, as part of the constellation of systemic, symbolic and objective violence, on this utterly warped view of society.

The roots of Foucault's ideas can be found in Nietzsche's radical relativism, where the ordinary constraints of life are framed as illegitimate obstructions to the will of the *Übermensch*.

> To refrain from mutual injury, mutual violence, mutual exploitation, to equate one's will with that of another: this may in a certain rough sense become good manners between individuals if the conditions for it are present (namely if their strength and value standards are in fact similar and they both belong to *one* body). As soon as there is a desire to take this principle further, however, and if possible even as the *fundamental principle of society*, it at once reveals itself for what it is: as the will to the *denial* of life, as the principle of dissolution and decay. One has to think this matter thoroughly through to the bottom and resist all sentimental weakness: life itself is *essentially* appropriation, injury, overpowering of the strange and weaker, suppression, severity, imposition of one's own forms, incorporation, and at the least, putting it mildest, exploitation. (Nietzsche 1973, §259, his emphases)

Nietszche's intellectual descendants go astray when they take this description of what is open to the *Übermensch*, who is after all fiercely intense, superior, transgressive and excessive, as generally applicable. If one has to read Nietzsche to discover how the *Übermensch* must live, one is probably not one. I write this at my desk, fending off the last nagging traces of a cold, watching the drizzle fall onto a construction site, nibbling on a Twix, and to be perfectly frank, I have never felt less *übermenschlich* in my life. Yet if most of us are not *Übermenschen*, then we are ordinary Jills and Joes, and so the comforting good manners of trustworthy conduct start to look rather more attractive, even as the fundamental principle of society. They may be a prison that suppresses our nobility and will to power, but I am happy to swap these for the Twix, thank you very much.

CHOOSING ONE'S OWN CONSTITUTION: DISSENT, DISOBEDIENCE AND LAWBREAKING

The *Übermenschen* who eschew mundane trust and trustworthiness are more than keen to take on the responsibility of deciding what laws they need to follow. The constitution, that in theory dictates the terms of political interaction, can, in the opinion of many, simply be ignored if it is not to their taste (for an overview of different ideologies' attitudes to constitutionalism, Greally 2024).

Chile's constitution of 1980, which led to relative economic and social stability and success for the country in the decades following (Zitelmann 2019, 92–107), has been called a "cheating constitution" (Ansaldi & Pardo-Vergara 2020, 17–28), because it created a number of mechanisms to make reform difficult (as is standard constitutional practice). This supposedly cemented neoliberalism "against the people" (Ansaldi & Pardo-Vergara 2020, 30). Another way of putting this is that it protects the Chilean market-based economy against those who disapprove of it.

As it is, "the people" protested against it in 2019, and were given an opportunity to replace it with a "progressive" constitution, with 388 articles guaranteeing rights to neurodiversity, a decent old age, digital connection and disconnection, culturally appropriate food, sex education, sport, an uncosted state-run care system from birth to death, and the freedom to strike for any purpose (Rioja et al. 2022). Many of its clauses were vaguely drafted, and the whole would have been fiscally disastrous. The Chilean people rejected it 62%–38%, against the campaign of its government (a far left government that had somehow been elected under the 1980 constitution). Academics were astonished that such a progressive document might have been rejected by a giant majority, lamenting "the powerful sway neoliberalism and nationalism continue to hold over Chilean politics and society" (Hiner 2022, 415; Escudero 2024).

"The people" also had an opportunity to vote on an extreme right constitution that was just as awful, and just as sensibly rejected that too. The net result of this three-year wrestling match which exhausted the nation's political actors was that the Pinochet-era constitution remains in place (at the time of writing). Few seem to have appetite for another attempt, but who knows? A process born of mistrust of politicians, and failure to appreciate the positive aspects of Chile's post-Pinochet history, was a colossal waste of time.

According to Ansaldi and Pardo-Vergara (2020), the 1980 constitution, because it cheats (i.e. it renders radical change harder to implement), is not a constitution at all, but rather a set of constitutional laws designed to skew the playing field. This self-serving distinction is taken from Schmitt (2008), whose agonism is unaccountably popular with progressives. This allows the critical left, convinced of its monopoly of moral rectitude, to reserve the right to cheat a cheating constitution. Castells defended the right of Catalan President Carles Puigdemont to hold an illegal referendum on independence, as he was running "a peaceful political and institutional movement which does not stay within the confines of the constitution, precisely because this is what the movement is questioning" (Castells 2019, 117). He warned that Spain "is on the edge of a constitutional precipice, which could jeopardize the very fabric of civic coexistence" (2019, 124), rather missing the point that not respecting a common set of rules provides little basis for trust, and pushes everyone closer to that precipice (Oakeshott 1975, 128). Indeed, playing by the rules is itself a problem for Castells: "activists are faced with the same challenge and dilemma …: how to take part in institutions and change politics without being co-opted by the system" (2019, 125). Merely acting democratically and respecting one's opponents is in and of itself an obstacle to the radical politics Castells espoused. But this is of a piece with his view that chaos is better than order you don't like.

The idea of choosing one's own laws, even in a democracy, has long been a principle of the activist left, from hunt saboteurs (Tichelar 2017, 147–163), to animal rights campaigners (Monaghan 1997, 2013), to those opposed to genetically engineered foods (Schurman & Monro 2010), who are pleased to apply thuggish techniques of dissuasion and deterrence against those whose perfectly lawful activities they find offensive (Baier 1994i is surprisingly ill-judged on this topic). Those wishing to take "direct" or illegal action against democratically legitimate policies or institutions cite the irreversibility of some policies (e.g. in the environmental sphere) as the justification for "having" to take undemocratic action: "democratic procedures fail to guarantee the realisation of crucial ecological values, and can thus be unjustified, even though legitimate" (Humphrey 2006, 314). Environmentalists, and others on the left who promote substantive values, should therefore only support political processes that are appropriate for realising those values in the world (Goodin 1992; Westra 1998). The same arguments, it might be said, could also be embraced by activists on the right, who (for instance) might assert that because large-scale immigration is irreversible and undemocratic, violent protest against it is justified. The obvious point is that, if my friends (to use Schmitt's terms) can choose which rules they follow, so can my enemies, unless we have enough violent methods to hand to prevent them (or *vice versa*). Not a good basis for a functioning society, although it might please the agonistically inclined.

However this goes, the critical left's rejection of its obligations to the machinery of democratic politics fundamentally undermines trust in the smooth running of society (or its systemic violence, as Žižek would have it); one would never know when a thug with a pot of paint or half a dozen eggs would be waiting round the corner. Although the critical left's attitude exceeds by a margin the liberals' advocacy of vigilance and dissent (Rosanvallon 2008;

Urbinati 2015, 180–182), it is continuous with it. Each position discounts any value provided by a well-run society, and instead privileges the claims of individuals to make their own rules, and run their own experiments with living. Where they differ is that the liberal cherishes the capacity to disagree without breaking the law, but, if determining our own way of life is our central right, it would be unsurprising if many people didn't wonder why they should obey democratically-established law when it conflicts with their own instincts.

This puts the last nail in the coffin of the liberal choicelessness theory that a well-written constitution can curtail authoritarianism and engender political trust. The converse thought is more useful: *if* political trust is widespread, a constitution *will* help coordinate political behaviour, but *isn't* necessary for it. Convention will do the same job.

CONCLUSION: TRUTH AND TRUST

The radical/progressive/critical tradition following Nietzsche is disastrous for trust. Nietzsche himself thought that trust and trustworthiness were incompatible with freedom and authenticity. His political project decoupled honesty and truth: she who behaves honestly in her devotion to her own autonomy, experience and will, will have no compunction in ignoring the truth. "Truth is ugly. We possess art lest we perish from the truth" (Nietzsche 1967, §822). Truth-telling, on this account, is dishonesty, an inauthenticity we are forced into when we are too weak to assert ourselves and resist the power of others.

It cannot be emphasised too much how far this obsession with power as the chief determinant and corruptor of human relations undermines trust. We can understand trust in terms of power – trusting someone may impose responsibilities on her, or the trustor may become dependent on her, while mistrusting someone may deny her an opportunity or be an injustice (Hawley 2017, 69–70). But these relations certainly don't exhaust the options, and not every episode of trust, mistrust, trustworthiness and untrustworthiness is seriously influenced by power (I would argue only a tiny minority are).

Not only does the obsession create what seems to be an unbridgeable gap between trustor and trustee, it also undermines the contributions that truth, reason, rationality and knowledge make to a congenial, civilised society. For Bourdieu, the presentation of the truth involves symbolic power, an exercise in symbolic violence. For Foucault, science and knowledge make no sense if we try to understand them outside their complex nexus with power (power/knowledge, in his terms).

Theorists of trust have often correctly emphasised the connections between trustworthiness and truth-telling (Hawley 2019, 48–71). "A politics that works within the medium of speech will, owing to the internal logic of language use, generate trust as a condition of communication. … [The recognition of the other person as a speaker] is an initial extension of trust that a speaker will say things that he/she considers true, sincere, or normatively right" (Warren 1999, 342). Irony, the mark of the Nietzschean *Übermensch*, is inimical to trustworthiness (Reemtsma 2012, 22). Furthermore, a connection between truth, trust and trustworthiness is what enables trust to produce positive social effects through social interconnection. "Truth is the medium which acts as carrier for the reduction of intersubjective complexity. Trust is only possible where truth is possible, where people can reach agreement about any given entity which is binding upon a third party. Truth facilitates this understanding, and thus the reduction in complexity, by the assumption that the third party would also consider their view as correct" (Luhmann 2017, 57).

I would add, pragmatically rather than epistemologically, that there is an awful lot of agreement about how the world stands (not just factually, but socially and morally), and that there is a great deal of trust around the place, despite a half-century campaign by the critical left to undermine it. It is evident to anyone with eyes to see that their assessment of the role of power in human relations is a gross overestimation. Their attempts to problematise truth, reason and broad, if shallow, consensus about goodness, virtue and social positivity have done little more than confuse and irritate, spawning over-defensive reactions, insularity and at worst right-wing populism. Once more, the Puritans enable the Peasants.

12. Identity politics against trust

The final brand of blockchain politics to be covered is identity politics, a somewhat imprecise term (Kurzwelly et al. 2023) denoting a style of politics based around groups and their perceptions of oppression. The private sphere, interaction, expression and even thought itself are policed, reducing discretion, so that trustless trust becomes the norm; behaviour that does not serve the interests of oppressed groups is outlawed. Recall the second tenet of blockchain politics – one can only trust a representative who has identical interests. The group is generally defined by a common ascriptive identity.

This last condition is surprisingly difficult to work out. I tend to sympathise with Laurance Splitter's concern that:

> … the dominant view of identity in the social sciences and media (including the idea of "identity politics"), is both confused and dangerous. Confused, in that it fails to distinguish strict (quantitative, numerical) identity (which makes an object or person the very same thing over time) from mere qualitative identity or similarity (which allows objects and persons to change over time); dangerous because it leads to a view of persons as overly dependent on the groups and institutions—nations, religions, cultures, ethnicities, gangs, cults—with which they often identify (or are often identified with). (Splitter 2022, 2)

It is tricky to decide which qualities are identifying, and which not. Identity politics tends to revolve around ethnocultural heritage, nationality, religion, sex and gender, and sexuality, although it is constantly expanding, incorporating qualities to do with medicalisation and disability, weight, age and generation, and caste. Some with attention deficit hyperactivity disorder (ADHD) have seen themselves in this way. "Rather than being an omen of difficulties to come, for many of us it's amazing news. … A diagnosis is the thing that allowed us to forgive ourselves and finally start healing. We are forming a robust, self-determining community for the first time" (Boseley 2023). Some want to include class identities, to align it more closely with the Marxist tradition, but this has proved harder than its advocates would have liked, so that many on the left see it as a dead end (Neiman 2023). Nevertheless, job adverts in the arts sector encourage applications from those who identify as working class, benefits class, criminal class or underclass (Malik 2024).

When identities are the basis of one's treatment in the world, one is highly incentivised to pick a side. One may decide to "identify" with a group that one is not obviously part of (Brubaker 2016). The indigenous people of Canada (Inuit, Métis and First Nations) are offered certain rights and land in the constitution, incentivising identity arbitrage. The Eastern Métis, whose claim to be legitimate Métis is rejected by Métis representatives, are growing rapidly "due to white Franco-Québécois and Acadian settlers 'becoming' Indigenous" (Leroux & Gaudry 2017). Other "indigenous peoples" have appeared almost overnight (Cecco 2024), or have switched allegiances for best advantage (Cecco 2021). The Canadians even have a word for it: Pretendian.

Identity goes beyond connections with the little platoons and associations which comprise a contented community; it requires stronger ties with perceived normative force. Conservatives and critical identity theorists share suspicion of universal principles (for example, the leading critical race theorist Derrick Bell rejected the campaign for school desegregation because he believed, entirely reasonably, that it distracted from the concerns of black parents about the quality of their children's education – Bell 2004). But the conservative also points out that the flexibility, cooperation and camaraderie characteristic of voluntary associations, which help communities to rub along together, are often absent from qualitative identities, which provoke more antagonistic relations.

Identity politics is a mainstay, regrettably, of the modern scene. It is related to critical progressivism, sharing the assumption that power, domination and oppression are ubiquitous in, and characteristic of, the happy, prosperous, free, relaxed democracies in which we are privileged to live. On Mounk's hostile account, the relatively austere invective of academic identity theorists was coarsened by activist journalism, whose simplistic *précis* are more successfully propagated on social media (2023, 84–96); I don't dispute his reconstruction, but I'm not sure the journalistic memes misrepresented the basically simple ideas behind identity politics (Pluckrose & Lindsay 2020 are less inclined to be sympathetic to activism masquerading as scholarship).

Identity politics has been connected with the label "woke," which is not helpful in formal discussion. It is also strongly connected with the Puritan tendency in politics. Susan Neiman introduced the topic by emphasising how generous left-wing urges tend to be subsumed by bad effects. "It begins with concern for marginalized persons, and ends by reducing each to the prism of her marginalization" (Neiman 2023, 5), but this overestimates its well-meaning nature, and understates its confusions. Rather, it often begins by taking a construct that is spurious, or in the eye of the beholder, and arguing that it marginalises those classified in that way.

Hence "black" has been defined in many inconsistent ways through history, usually to the detriment of those so classified. One can reject such labels as inadequate to capture reality (Williams 2019), or be indifferent to them, becoming, if not colour-blind, at least what Angel Eduardo has called "color-blah" (Loury et al. 2024). The identity politician's clever wheeze is, rather than to criticise the classification, to take it as "real" and use it to form a resistance group to the oppressor (Spivak 2006). This is not the intellectual cop-out that some have implied (Mounk 2023, 44–47), since we occupy the realm of appearance: "it is on the surface that we live and act: it is there that we are created, as complex appearances sustained by the social interaction which we, as appearances, also create" (Scruton 2016, 10).

However, the classification is reified by identity theorists beyond such tactical essentialism, because it is taken to determine the ways in which we understand the world. Hence, one's identity, particularly if one is oppressed because of it, means that one *necessarily* sees and experiences the world differently from the oppressor. This, understandably, problematises dialogue between oppressed and oppressor.

Hence, while identity politics is unconcerned with contingent bonds of association, and focused on identity, that distinction is rendered arbitrary, since it depends on the subjective experience of marginalisation. Identity implicates the sense of self, but as seen solipsistically from the inside. Critical identity theory tends to be fairly hostile to little platoons, seeing threats to its take on identity from traditional associations like the family, regiment and firm (Bell 1980, 524–525; Flaskas & Humphreys 1993; Harkins 2012; Taylor 2012).

There is clearly something here about self-esteem and dignity, which resonates with our discussion about community, legibility and commonality. If you are infected with the Foucauldian virus, then you will think about these in terms of power, about how the minority is dominated by a powerful majority, and needs to exercise its own will to power. This will make it harder, not easier, to join a community and enjoy its benefits. However, perhaps you already think that its benefits are denied to you.

Given its sociological roots, identity politics sees domination as embedded in social structures that are perpetuated and reproduced by quite ordinary and apparently innocuous interactions. It cannot therefore be eradicated by applying colour-blind or unisex rules indiscriminately, because the ability to follow the rules at all is impacted by previous experiences and path dependencies. The conservative tends to agree with the identity theorist about this, but does not see anything illegitimate going on.

Identity is rarely monolithic – one has a race, a sex, a nationality, a religion, and possibly more than one of these. This led to the complication of *intersectionality* (Cho et al;. 2013 Crenshaw 2017), which might have been a commonsense watering-down of the theory, but instead ratcheted it up. The oppression of a black woman is not a simple function of the racism and misogyny she experiences, but instead is a complex matrix that multiplies their force. As identity politics is inherently oppositional, and as intersectionality effectively multiplies identities and reduces the numbers in the groups they subtend (ultimately down to uniqueness), we get closer to a war of all against all, a point emphasised by traditional Marxists (Das 2023, 22–24). For instance, white feminists become a target of their non-white (former) sisters.

> Chief among [white feminism's core mechanics] ... is the acute ability to craft contemporary images that tell a story of progress ... while maintaining power structures as is. This dangerous maneuver allows white feminism to usurp the accolades, scholarship, efforts, and knowledge of people of color, of queer people, of disabled people, of all disenfranchised people and use it against them within the very institutions they hope to change. (Beck 2021, 24)

Intersectionality *per se* will only produce tractable outcomes if there is an artificial limit to the intersections that are deemed appropriate – in other words, there is a limited number of things that can be attached to one's identity. Otherwise, the intersectional urge will continue to both lift and separate, as our associations, contextualisations, connections, specific standpoints and anything else deemed relevant will ultimately divide us into monads.

Conservatives have long warned about identity politics as antagonistic, preferring the careful construction and curation of national, community or cultural identities (e.g. Buckley 2018, 57–58; Skelton 2019, 131–145; Timothy 2020, 101–154). In this chapter, I will extend this insight to consider the relation of identity and trust in more detail, including: the connection between identity and experience; the focus on victimhood; and the submersion of the individual into a group identity.

IDENTITY AND EXPERIENCE

To begin with, I will look at a major driver of the identity theorists' normative programme: the identity-mediated experience of the social world. The basic position is straightforward – those who do not experience oppression or domination can never understand or empathise with those who do.

Standpoint Epistemology and its Consequences

Standpoint epistemology (Toole 2022) is the name given to the view that someone's identity is a crucial factor in understanding how they come to experience the world and gain knowledge about it, and put like that is a reasonable point to make. Those at the sharp end know what it's like to be at the sharp end. It contrasts with a purer epistemology which ignores non-epistemic features such as the social status of the knower. It has augmented feminist theories that focus on power relations between the sexes (Collins 2017, 119–120). It was never intended to deliver a theory of truth, not least because its Foucauldian intellectual roots implied that there could be no such thing.

However, it privileges factors of identity over other contextual factors. Certainly a professor of black feminism will see her race and sex as central to her experience of the world, because that is her job and vocation, but one wonders how exportable her standpoint is to the rest of that intersectional group, who may see their identity through wider lenses, if they consider their identities at all.

The importance of standpoint has long been recognised, if not in mainstream epistemology. Quine argued that epistemology should take seriously the characteristics of the body's sensory apparatus, which provides a basic filter of what evidence can be uncovered, and therefore what can be known (Quine 1969), while Nagel stressed that a human biologist could know everything there was to know about bats, except what it was like to be one (Nagel 1974). A distinction is commonly made between *Erlebnis*, the private or subjective perception of lived experience, and *Ehrfahrung*, experience that can be shared with others (Splitter 2022, 142–148). It is hard to imagine how to share lived experience – can I *really* understand, except in very broad terms, what it is like to be, for example, a woman, a skilled musician, a brave soldier, bilingual, short, devout, wealthy, schizophrenic, or fiercely competitive, when I am in reality none of those things?

The problem that intersectionality poses for standpoint epistemology is that it prevents the resolution of this perspectival problem. If an oppressed social group has *Erlebnis* in common that is denied to outsiders, how can they be sure that what is *Erlebnis* relative to outsiders is *Ehrfahrung* relative to each other, i.e. that they can share their lived experience? Women may have different lived experience from men, but do white women and black women? Do mothers and the childless? Do heterosexuals and lesbians? Do biological women and trans women?

This isn't a sticking point for the rest of us. We saw in Chapter 7 that there is something to common understandings of the world, for example helping community coherence, while erecting barriers to entry. Yet in that view, there is not a divide but a continuum – there is nothing *stopping* outsiders getting to understand a community, and nothing to prevent common experiences developing to bring separate groups together. None of that need implicate identity at all, and Chapter 7 makes no claim that community members and outsiders have different identities, and certainly not mutually exclusive ones.

The problem with standpoint epistemology is that it requires an axis (such as race or gender) to separate *Ehrfahrung* from *Erlebnis*. Lived experience of, say, black women, is something that can be shared and used to understand how they are oppressed, and maybe do something about it. However, it cannot be transferred to black men, white women (even feminists), and certainly not white men. But equally, a middle class black woman from the same town as me, or who has had a similar career, may well be closer to me in important respects than to a farmer's wife in Northern Nigeria, or a factory worker in Georgia. A single father would probably be pretty close to a mother in his appreciation of the demands of childcare.

It seems far more reasonable to make the weaker claim that a white man such as myself will struggle to understand what life is like for a black woman, depending on how far the ranges of our experiences coincide. Conversely, it doesn't take a deep empathy based on lived experience of prejudice to reject racism. I may be more or less sensitive in my appreciation, and I may succeed or fail in connecting across the divide, but there is no reason to suggest in advance that it cannot be done.

Incommensurable Lifeworlds

Policing the *Erlebnis/Ehrfahrung* boundary produces perverse effects. If we assume that different standpoints prevent understanding, then communication, civility, trust and public life must break down.

For instance, philosopher Jonathan Wolff argued for diversifying university curriculums to reflect the diverse student body.

> In my own case I teach political philosophy, typically to students from more than 50 countries. Every year my colleagues and I comb over sources to find accessible non-white, non-western authors on the topics we teach. It's a struggle. What can you do if you can't find the texts? This year we are taking what now seems the obvious step of considering changing some of the topics, too, such as including African ideas of community and nationhood, as well as the Black Lives Matter campaign. (Wolff 2020)

This is nothing short of weird. The course on political philosophy, at Oxford, was presumably originally focused on the charismatic megafauna of Western ideas – Machiavelli, Hobbes, Marx, etc. It attracted students from 50 countries, keen to be taught by Wolff and his colleagues. Yet Wolff took this as a signal, not that they were providing a service that was in demand, but that students from countries or cultures which aren't white and Western will struggle to understand, or be bored by, or be repelled by, or otherwise be unfitted to study, the dead white males. "This privilege must end," thundered he, proposing to end it by teaching Afro-communitarian philosophy (certainly an interesting and enlightening topic), a subject about which he knows nothing, or at least not much more than his students, and about which they could learn far more at London's School of Oriental and African Studies (SOAS), never mind one of the many high-quality universities on the African continent.

As another example, Labour MP David Lammy complained that the inquiry into the appalling Grenfell Tower fire in London was being chaired by a white judge who could not be trusted by the victims' families because he looked too much like the builders, contractors and regulators under investigation (Timothy 2020, 128–129). The judge was an elderly male lawyer – but more pertinent was his experience and ability to orchestrate the interrogation to extract the important information. If what we want from an inquiry is an exploration of responsibility and regulation, then we require legal expertise, and trust (not unconditional) in the operation of the law. If one cannot trust a legal process conducted by someone unlike oneself, trust in the conduct of the law is not justified, and there is no ground for holding an inquiry at all. Lammy, a lawyer educated at SOAS and Harvard, might have rather more in common with the judge than with some of the undocumented migrants and asylum seekers living in extreme poverty in the building; he didn't make clear why he was able to represent their views, while the judge was unable to conduct the inquiry.

The incommensurability thesis reaches its zenith with the doctrine of *white fragility*, that white people cannot escape their privileges and get defensive about them when challenged; thus they are unable to be effective anti-racists (Thompson 2003). "White equilibrium is a cocoon of racial comfort, centrality, superiority, entitlement, racial apathy, and obliviousness, all rooted in an identity of being good people free of racism. Challenging this cocoon throws off our racial balance. Because being racially off balance is so rare, we have not had to build the capacity to sustain the discomfort. Thus, whites find these challenges unbearable and want them to stop" (DiAngelo 2018, 112). Much anti-racism is therefore parasitic on racism itself; whites, on this reading, cannot be trusted in the struggle against racism (both authors cited, it should be stated, are white American women).

When lifeworlds are taken to be radically incommensurable in this way, public life is inevitably affected.

> Without a shared universe of expectations, histories, memories, or affective commitments, no basis of familiarity can exist. In a situation of radically incommensurate life-worlds … that familiarity necessary to negotiate role expectations … is lacking, and beginning to emerge in its place are the increasingly public definitions of roles and role expectations … . In the absence of trust, indeterminacy becomes intolerable, hence the continued promulgation of "speech codes," housing association regulations, smoking laws, and other forms of formal regulation (and sanctions) of interpersonal behaviour. (Seligman 1997, 154)

Trust cannot function, because would-be trustors cannot fathom what would-be trustees are going to do, and so we need to regulate – moving closer to trustless trust.

The logic of the argument separates and estranges. If diversity tends to be negatively correlated with solidarity, then, in a diverse world, solidarity may suffer (Timothy 2020, 58–59). There is a slippery slope from decreased trust due to diversity, to decreased trust due to politicisation of ethnicity, to decreased trust due to multicultural separatism (Lenard 2012, 97).

Allyship

If every conceivable intersectional group is mustering to defend its interests and combat the domination it faces without compromise, dialogue or trust, an obvious question is what the chief oppressors, white heterosexual males, are going to do. They have two options, allyship or embracing their own identity.

The former is a dead end. Allyship was a predictable, if recent, development of critical identity politics, pronounced Word of the Year in 2021 by a dictionary that had only added it a month previously.[1] Given that racism is structural, systemic and omnipresent, then no white person can escape it – merely behaving well and equitably cannot be enough (Roberson 2024). Racism isn't committed by racists, but is perpetuated or reproduced by the functioning of white society. This implicates all white people in racism, a line that suits anti-racists of all colours, emphasising the depth of social change needed.

However, it puts white anti-racists in the embarrassing position of being no different from Donald Trump. Hence the ally *had* to be invented: someone who accepts the fact and shame of their white privilege, expects no favourable regard from people of colour, and instead uses their privilege to try to dismantle racist systems – thereby achieving an intermediate status. Allyship is like Premium Economy on an aeroplane, way short of Business Class, but slightly more legroom than Economy.

In terms of trust, it is a deeply asymmetric situation. People of colour, on the theory, should never trust white people, because they cannot be trustworthy (owing to systemic constraints). The ally, notwithstanding, aspires to be trustworthy, working to the benefit of people of colour while accepting that she will never gain their trust. "This notion, one of *true allyship*, extends so far beyond the purview of contemporary White engagement with racial justice that it seems fanciful; almost laughable" (Smoot 2020, their emphasis).

And so the ally creeps around, unloved and unlovable, Uriah Heep without the saving grace of malice. The following is a plea to a problem page; it is so bizarre that it may be made up or even a spoof, but it was presented as genuine.

> I'm a 60-year-old white American male and have done some work and had training around acknowledging my racist background and lingering internalised superiority.

---

1 https://www.dictionary.com/e/word-of-the-year-2021/.

> In multicultural environments, is there a way to indicate my allyship with other people, letting them know I'm safe, or at least not hostile, without coming across as trying too hard? My approach has been "act naturally" but I kind of want to let people around me know they can relax a little. Perhaps it's just a moment of eye contact and a head nod. And maybe this question reveals I'm not even ready to be that person. (Msimang 2023)

Certainly you would not want this fellow around, whatever your race, leering benignly at you.

Owning Your Prejudice

If allyship won't work, the alternative for the oppressor is to embrace his identity without shame. White males will have no difficulty in demonstrating how *they* are oppressed. Everyone is shouting at them, masculinity is widely held to be toxic and the patriarchy responsible for every evil (Power 2022, 15–93). Boys find it hard to navigate a world in which they are constantly being lectured for what appears to be original sin (Carr 2024). Men have lower levels of health and education, and their physical strength is no longer an advantage in the digital economy.

Generations of boys and girls have been exposed to a broad feminist consensus in education, health and social services about toxic masculinity. Yet – surprising to the educationalists, less so to anyone else – boys are flocking to the ravings of social media "manfluencers" like Andrew Tate (Botto & Gottzén 2024; Wescott et al. 2024), while record numbers of girls want to become boys (Barnes 2023, 97–98; Cass 2024, 85–86). The manfluencers lurk in an online space called the "manosphere" (Ging 2019), characterised by "hybrid masculinities," such as involuntary celibates, beta males, sigma males, misogynists, geeks, wellness bros, pick-up artists, red pillers, betafags, etc., basically a loose, largely online network of men cross about feminism and concerned that they are being systematically suppressed or even wiped out. This, like the race-based great replacement theory (Cosentino 2020, 59–86), is obviously mad, but some in the manosphere are driven, or drive themselves, to violence and occasionally terrorism. Even if they are non-violent, it's not great that a lot of men think mad things.

Meanwhile, "whiteness" is an omnipresent but hard-to-detect force oppressing non-white people, normalising white racial domination and tainting virtually all white people with guilt (Ahmed 2007; Ferber 2007; Garner 2007; Nayak 2007; Withers 2017). If enough people tell you to "own your racism" (DiAngelo 2018, 139–147; Korducki 2020; Jackson & Rao 2022), then you might begin to think that doing so and living with it is the path of least resistance, since "a positive white identity is an impossible goal" (DiAngelo 2018,

149). It is an unfortunate quirk of critical identity theory that the very idea of white people attacking and dispelling the evil of racism becomes unthinkable. The argument that "the interests of blacks in achieving racial equality will be accommodated only when it converges with the interests of whites" (Bell 1980, 523) is probably true. However, the tactical question of whether confrontation or compromise will achieve the desired change in whites' understanding of their own interests is moot. Confrontation seems to be producing pushback, not retreat.

Critical studies of social oppression, such as critical race theory, seem only able to say that we cannot live together. Hence their only possible outcome, given that we do have to live contiguously, is that society will stratify as each accepted identity develops organisations and institutions to defend it from the others. Consequently, the Peasants play the game themselves, and, given the head start everyone else had, they are not doing badly. Some influencers with ghastly views have millions of followers, and populists of all stripes are increasing their vote share across the democratic world.

The Puritans persist in the game. Apparently, some schools in America segregate the races in class, to affirm, and sometimes create, children's racial identity (Mounk 2023, 1–4). The parallels with the arguments of the *apartheid* regime are uncomfortable to say the least.

VICTIMHOOD

Victimhood, which used to be a stigma, is a source of standing in identity politics (Neiman 2023, 16–17). One's achievements matter less than one's suffering – and in many cases the suffering of one's ancestors, or those one identifies with. We have already seen the manosphere as a locus for men's claims to victimhood, and it has been suggested that victimhood is now a common strategy for relatively well-off majorities (Aronson 2021).

This creates a problem for trust. If the way to get on is to be a victim, then trusting will always be a mistake. A victim may have trusted in advance of his victimhood (the betrayal of trust may be a crucial aspect of it), but certainly cannot trust afterwards. All he can do is denounce. Trust, in identity politics, can only be shared with those who are like oneself, and – given the demands of intersectionality – even then only partially.

Trust as a Losing Strategy

Trust in the identity world is a losing strategy. Trust of an oppressor exhibits a range of virtues that ameliorate valuable victimhood – sympathy, faith, charity, stoicism – all felt to be suspicious by identity theorists. This was amusingly exemplified in, of all things, the report of a ladies' tennis match in *The*

Guardian, which announced that the "concept of being a 'good loser' …, supposedly tied up in ideas of sportsmanship and fair play, is the sort of moralistic crypto-imperial nonsense that late-Victorian Britain seemed to churn out like ore" (Liew 2023). Another columnist in the same newspaper was baffled when kind and thoughtful friends in Ghana sent her their condolences upon the death of Prince Philip: "the unspoken requirement for us to publicly celebrate the monarchy's gains – or mourn any of its losses – demands that I internalise a history of violence and racism against my own ancestors" (Hirsch 2021).

Political status and power follow from the contrary vices: anger, mistrust, disbelief, oversensitivity. The aim is to sabotage trust in trust, what Luhmann called reflexive trust (2017, 73). Each group closes down trust and trustworthiness with respect to outsiders, because not to do so is to miss opportunities for political advancement.

At a personal level, one must "call out" bad behaviour and microaggression. The lack of intention to offend cannot be taken into account; the ability to take offence is an opportunity, and one would be irrational not to take it if it was offered. "I was brought up to believe that you should never give offence if you can avoid it; the new culture tells us that you should always take offence if you can" (Scruton 2022, 85). Calling out has to be public, to shame the offender, and register the offence. Hence, the "calling out" culture is a win-win in identity politics, as it demonstrates victimhood, while also tending to make the person called out defensive and prickly (the fragility of privilege), thereby poisoning future relations (DiAngelo 2018). In that way, both racism and identity politics are reproduced structurally.

Politeness, agreeableness and the other Aristotelian civic virtues are, unfortunately, too constructive.

> The virtues of politeness are a prelude to overcoming … misunderstanding because the polite person is one who understands that she is not the centre of the universe and that cultural sensibilities need to be cultivated. Such a person is informed by a habitual disposition to be open to others, to foster the conditions of trust, to performative sensitivities, and as such is a conduit, within differing social contexts, for the flourishing of deeper attachments. The virtue of politeness is also especially valuable when there is radical disagreement. (Mooney & Roy 2021, 192)

Civility, politeness and kindness are stigmatised in critical identity theory as perpetuating racism and sexism (Bennett et al;. 2013 Zamalin 2021; Jackson & Rao 2022). They do not challenge; silence is violence.

Calling out is not, however, unproblematic for identity politics. In the first place, as a response to specific events and incidents, it tends to suggest racism consists primarily of momentary pieces of abhorrent behaviour, but this goes against critical identity theory doctrine (Bell 1993; Bouvier & Machin 2021; Woods & Ruscher 2021; Kim et al. 2022). Hence the identity theorist has to maintain that these microaggressions are just the visible tip of the structural iceberg. Calling out has also led some Marxists to be suspicious that identity theory is chiefly a means for ambitious members of minorities to shame their bourgeois colleagues into promoting them. "The slightly better-off among oppressed groups, for whatever reason (including pure chance), compete for the top posts with those of culturally dominant groups; so they use [identity politics] to meet their needs and ambition. This suggests that IP as an idea becomes an ideology of these better-off sub-groups" (Das 2023, 27).

Getting Things Done

At the social level, the disdain for trust certainly helps with the name-calling and oneupmanship that characterises beggar-thy-neighbour politics, but it is not conducive to getting-things-done politics. An example that illustrates this is the Dutch BIJ1 party (pronounced Bye-ayne, a pun on the Dutch word "bijeen" which ironically means "together"). It was founded in 2016 as a self-consciously intersectional radical left party advocating system change to eradicate racism, sexism and ableism, and conceives of the Netherlands as an oppressive, racist society that can't be reformed incrementally (Seawell 2021).

The problem is that when calling out oppression is the only club in your bag, then you are more likely to get trapped in the bunker than to get anything in the hole. BIJ1 barely functions (as would be predicted by traditional Marxists, e.g. Das 2023, 25). Councillors, candidates and officials have been suspended for manipulative behaviour, sowing division and being masculine-dominant; others have quit, complaining of anti-black racism, misogynoir behaviour (i.e. misogynistic behaviour toward black women), Afrophobia, anti-pregnancy discrimination, and a toxic, structurally unsafe working environment. One official, forced out by a members' revolt, grumbled that "I've seen members want justice until a just decision doesn't suit them. That members would like a flat organization, but never want to be responsible for their own behavior and so on." The party's founder and leader quit in advance of the 2023 general election on grounds of health, adding that she was fed up defending her leadership against false allegations from party members. BIJ1 lost its only seat, as its vote share halved from the previous election in 2021. To put the tin hat on it, it had to repay €127,000 in subsidies after failing to provide documents to show they had been spent properly.[2]

You couldn't make it up: the only specifically anti-racist, anti-sexist party collapsing in mutual accusations of bullying, racism and sexism. The

[2] For newspaper articles detailing these various events, see https://nltimes.nl/tags/bij1.

Netherlands is a pretty tolerant and safe environment, all in all, and I have no doubt that BIJ1 itself is probably about as safe a workplace for women and ethnic minorities as you will find in politics. If you feel unsafe there, then you are not going to feel safe anywhere. Identity politicians, though, are not corporate beings, they work through aggression and mistrust to eschew compromise. That will lead to internal conflict, discord and personal affront, down less to discrimination than to the praeternatural ability to detect discrimination in the most tolerant of places.

Rainbow Politics

This leads us to the fallacy of the rainbow. Rainbow politics envisages a group of ideologically separate parties or voting blocs working together either to combat a particular force (e.g. capitalism or the Republicans), or to campaign on a topic that concerns a heterogeneous set of groups (e.g. climate change). Members of rainbow coalitions like to point out their numerical superiority in aggregate, and also the warm, fuzzy feeling they get working across political divides in a common cause (e.g. Thompson et al. 2022). In the US, the "Democratic coalition" was meant to put the Republicans permanently out of office (Judis & Teixeira 2002), though it seems to be fraying (Judis & Teixeira 2023). In Britain, Neal Lawson is a serial promoter of a "progressive alliance" to rid the nation of the hated Tories (2017, 2021a, 2021b, 2023). In the end, the hated Tories self-detonated, and the alliance wasn't needed.

The metaphor is meant to convey the beauty and contrasting colours of a celestial rainbow. It is appropriate, however, not for this reason, but because a rainbow is an arc of formerly coherent white light that has been made incoherent and disjoint by a prism. Any coherence a rainbow coalition has depends on what it opposes. It is noticeable that, in an ideology that claims not to like binaries, rainbow coalitions work by specifying a unifying binary. "People of colour" is a great phrase, somehow excluding only white people (white presumably not being counted as a colour) and co-opting everyone else, allowing them to be characterised as the "global majority." But is that an identity? Lesbian and gay activists did the same by the clever means of adding "+" to an acronym (it began as "LGB," and there were many debates about what new letters might be allowable and what they might stand for). Now, instead of having to reconstruct the acronym every time a new sexuality surfaced, anyone who doesn't feel particularly attached to mainstream heterosexuality (or cisnormativity as it is now called) is added in, whether they wish to be or not. Coalitions are hard to maintain in the world of identity politics, but they are easy to announce.

The result is not harmony within the rainbow, as BIJ1 shows. Political battles between, say, Latinos, black Americans and Asian Americans (e.g.

Roberson 2024, 7–8), who may or may not identity as people of colour even as they are drafted into the rainbow, or between transgender people and lesbians or gay men (Roberts 2024), testify to the malnourished level of trust within the wider groups. Meanwhile, the binary identity, such as "person of colour" and "LGBTQ+," sets up an automatic antagonistic mistrust of the supposed dominant oppressor, white people, including Jews, in the first instance, and heterosexual people in the second. Given the thin level of trust within, and its absence without, the rainbow is a fine symbol of chaotic war of all against all.

THE LOSS OF THE INDIVIDUAL IN GROUP IDENTITY

Identity has taken on a weight of meaning that it doesn't really deserve. Certainly people have attributes, some ascriptive, some adopted, that they care about more or less, but it is a new practice to corral them into those enclosures. "Until the middle of the twentieth century, no one who was asked about a person's identity would have mentioned race, sex, class, nationality, region or religion. … Identity [was] utterly particular and personal" (Appiah 2018, 3). My whiteness or maleness don't feel identifying, although I am, for good or ill, both those things, and no going back. From the point of view of being "singled out" in a crowd, virtually any piece of information can be "identifying" in that respect, which is what makes deanonymising personal data so difficult, a topic I have researched in some detail (Elliot et al. 2018).

Today's standard identifiers determine nothing. Given my social status based on my race, religion, sex, etc., in any situation it would still be open to me to behave in a stereotypical or an unexpected way. It is unsurprising therefore that supposedly monolithic blocs easily break up as their members' heterogeneity takes over and more personal interests (or apathy) appear more salient to them (Mounk 2022, 254). Consider the number of Trump voters who theoretically shouldn't vote for him (Lenard 2012, 103; Petty et al. 2022).

Unity Through Resentment (E Ira Unum)

Seligman suggested that this style of politics pushes out trust and replaces it with the pre-modern idea of solidarity, which is inimical to individualism (1997, 155, 168). Certainly a politicised group will make claims on individuals, so that a key factor for identity politics, therefore, is leadership.

> But human identities are deeply malleable. They take form – and change nature – in keeping with the narratives imparted by an older generation, the cues given by elites, and the incentives created by institutions. A lot thus depends on whether elementary school teachers and college professors, senators and CEOs help to support the natural processes that are making boundaries between different groups more porous … … [I]n more and more spheres of American life, well-intentioned

> people who genuinely believe that they are fighting for righteous causes are doing what they can to make racial identity the all-encompassing dividing line of American life. An overwhelming focus on the importance of ethnic identity and the irreconcilable conflicts between whites and "people of color" is quickly becoming part of the ruling ideology of the American elite. (Mounk 2022, 249–250)

To get political traction, such leadership involves not only making the case for group interests, but also stoking mistrust and animus. Identity politics is most corrosive where the educated congregate, for example in universities, cultural organisations and knowledge-based businesses ("it is Reason, not Authority, that is destructive of individuality" – Oakeshott 1991c, 282). University courses on gender studies, race and ethnic studies, queer studies, post-colonial theory, etc., together with rules of association drawn up by allies in the administration block, emphasise the ubiquity and invisibility of prejudice from untrustworthy dominant majorities (Pluckrose & Lindsay 2020), and "scholarship [is] the victim" (Paglia 2017, 83). "Critical pedagogy" (Freire 2017), with its implicit rejection of scientific epistemology and the possibility of a neutral stance, seems to devolve into an attempt to spread such ideas across the student body, while deliberately exempting itself from critique (Armstrong 2024). Indeed, few in the field seem to question whether the "pedagogy of the oppressed" is appropriate in successful, prosperous and liberal democracies at all.

A revolution in historiography and social science has shifted the spotlight of research. Traditional scholarship tended to focus on the people at the top, high politics, "great men," Whig history and grand narratives. In reaction, Marxists and others examined social and structural phenomena, class, the masses, "mass man,'" "history from below" and materialism. More recently, however, the trend has been to surface the marginalised, dispossessed, voiceless and oppressed. The key difference between the first two approaches and the third is that the last cannot exhibit any positive characteristics of any society under study. The first two may be sympathetic or unsympathetic to their subjects, and may find good at the top or at the bottom, but the relentless methodological focus on exclusion and oppression in the third approach must always miss the ways in which societies function and the benefits they create for their integrated members (Kekes 1998, 8–9). It is as if the highlights of a football match showed only the fouls; one could have no idea of whether one's team actually won.

Such history has turned our public institutions into penitential chambers of correction. In a splendid quote, a Tate Gallery curator justified yet another exhibition on colonialism and slavery by saying "If we can be proud of the best moments in history … we must reflect on the worst" (Economist 2023); but gave no hint about the point at which the reflection on the worst could stop, so that the Tate's pride in the best could finally be expressed. I was a member of the Tate for upwards of a decade, in which time not a single uncaveated positive description of Britain emerged, never mind pride. Trustworthiness, in this world, is signalled not via common reference points from the past, but rather by the eagerness with which one joins in the condemnation.

Whatever the historical wrongs, the groupthink of Tate Gallery reductivism serves oppressed or formerly oppressed people badly. For instance, one paper asserted without argument the pervasive influence of "the American cultural myth of an equal playing field in which exceptional social positions are thought to be earned through hard-work and talent rather than inherited through codified privilege" (Gimbrone et al. 2022, 7), as if this were an either/or. Surely social position is (partly) a function of work, talent and privilege as well as other stuff such as health, location, stability, family support, social capital, sobriety, good sense, amiability and dumb luck. You are dealt a hand, but you can still play it well or badly. The suggestion that privilege overrides *everything* is just another American cultural myth, but one tantamount to an attempt to sabotage the prospects of the underprivileged, by discouraging them from even trying. The harder someone works, and the more agreeable to and cooperative with others they are, the better they will generally get on, even if their lack of "codified privilege" simultaneously drags their prospects down. The American myth of the equal playing field may be fantasy, but it is more conducive to trust and collaboration than the equally false academic myth of the supremacy of inherited privilege.

Conflict of Group Interests

The uncompromising pursuit of group interests is an obstacle to the dissemination of trust across groups. Uslaner argued on the basis of survey evidence that tolerance of unpopular groups is the mark of the generalised trustor (2002, 179), but even he had to note that generalised trustors tend to come down against politicians who appeal primarily to minority communities (2002, 191, 197). At the very least, calling out and other politicising responses are off-putting for those who do not wish to cause offence but equally want to retain a modicum of freedom in what they say. So, far from promoting trust, identity politics fuels division and may even increase support for the far right. If *ressentiment* can be fostered within minorities, it can also be created in the majority. And – if they are the numerical majority – you may have lost the battle right there.

People will not defer to another group because told to by an elite, but will do if they think democratic principles are being challenged (Mounk 2022, 189). Conversely, the refusal to see a particular flashpoint in terms of basic rights makes every conflict zero sum (Deneen 2019, 29). For instance, the initial aims of the Black Lives Matter (BLM) movement were about rights, and nothing to do with collective guilt of white people (Neiman 2023, 27). Black people are disproportionately likely to be locked up, injured and killed by police officers in the US and elsewhere, a problem that should concern everyone. Yet BLM did descend into tribalism, identity politics, and unsupportable policy demands, not least defunding the police (Cobbina-Dungy & Jones-Brown 2023), a policy that was roundly rejected at the ballot box by all communities where violence and crime were perceived as threats. Even so, academics persisted in advocating defunding in other countries where the problem was far less urgent (Fleetwood & Lea 2022).

Identity-based policy almost inevitably depletes trust. The creation of the policy is likely to decrease minority groups' trust of outsiders, because it necessarily involves a framing of depravation, oppression and injustice. But equally, the implementation of the policy will typically result in losses for the majority that will be perceived as unjust by them. In one jaw-dropping example, there was a movement during the COVID pandemic to focus resources on minority groups, *even if this resulted in larger numbers of total deaths.*

> A strict MLS [Maximum Lives Saved] approach, while maximising total numbers saved, may also cause harms by disproportionately amplifying grief and mourning by communities of colour, and rightly justify increased distrust in medicine by people of colour. These consequences could not be classified as unintended; they would be known and anticipated. (Bruce & Tallman 2021, 209)

The (white) authors argued that it is not unusual to depart from the MLS standard, but the only examples they gave were discriminating in favour of pregnant women, first responders and people with disabilities, and removing older clinicians from direct duties to minimise their risk of infection. These would probably get support from expert and lay communities alike. The completely different and underspecified aim of trying not to "amplify grief," on the other hand, would spread mistrust through *all* communities. After all, no bereaved family from the majority could be sure that their loved one had not been sacrificed to the policy, while the bereaved in the minority communities would be more likely, given the rhetoric of mistrust and resentment that justified the policy, to assume it had not gone far enough.

The Totalitarian Instinct

Submerging the individual in the totality of the identity leads to many absurdities, but the biggest problem for trust is the totalitarian instinct to enforce the coherence of the group by fostering unanimous resentment of outsiders. This is intended to silence criticism from both within and without, enabling it to throw its political weight around. If all speech is symbolic violence, then free speech can have no special status (Timothy 2020, 41). It is bizarre, given the immense weight of evidence (Mounk 2023, 161–182; Norris 2023), that the existence of "cancel culture" is still denied.[3]

Where it is admitted, it is rationalised as "accountability." This may or may not be appropriate for public figures, but is exceptionally lacking in compassion when ordinary folk lose their livelihoods or public name over a minor solecism (particularly one which only the most sensitive of culture warriors could detect). Meanwhile, people and organisations are forced to take sides over issues over which they have no control, and which are irrelevant to their missions. Silence is violence – so that indifference to the oppression of marginalised groups is itself taken as an act of oppression. All sorts of utterly unrelated companies, charities and prominent individuals find themselves pressured to make statements about political issues that will have zero effect. Others are keen to make such statements, signalling virtue despite their irrelevance.

Foucault taught that civil society inevitably marginalises unfashionable or unpopular styles of life (Urbinati 2015, 162). Identity theorists have learned the lesson, and have become adept at the marginalisation business, at least in areas such as academe, the arts, the media and certain consumer goods and services markets, where they can project power by aggregating and exaggerating their voices via social media. The irony of the marginalised (or rather their vocal allies and self-appointed spokespeople) bullying their supposed oppressors is lost upon them.

CONCLUSION: IDENTITY POLITICS AND THE ANNIHILATION OF TRUST

The illiberal tendencies of identity politics and the iniquities of cancel culture have been well documented. What concerns us is the effects they have on trust, of which four stand out. The first is the most obvious: aggressive action by identity theorists toward those who are not persuaded by, or less committed to, their ideology simply puts up a barrier and makes empathy and trust less likely. If someone is called out or cancelled, especially when they do not believe they have been unreasonable, if their views are (or were until recently) more or less mainstream and shared with most in the community, the result will be more likely to be enmity than deference. Even if Bob succeeds in getting Alice cancelled, believing her to be racist, homophobic or misandrist, there is no reason to think she will change, and so from his point of view she will remain

---

[3] In the 2024 update of the Academic Freedom Index, produced by the Friedrich-Alexander-Universität, the UK had fallen to 66th place, the Netherlands to 64th, and the United States to 88th, below Senegal, Moldova and Kenya, https://academic-freedom-index.net/research/Academic_Freedom_Index_Update_2024.pdf, 10.

untrustworthy. It is merely that her voice is suppressed. Of course, this may increase trust in Alice from misandrists, homophobes and racists, and thereby strengthen their associations.

In identity theory, cancelling Alice holds her accountable for her views, and she will therefore learn, through shame, what is acceptable. She may even learn that she can never be acceptable at all, and live a perpetually wretched and contrite life as a humble ally (DiAngelo 2018). However, perhaps more likely is that Alice will cease to try to be trustworthy toward such an implacable audience, and simply exclude Bob from the intended beneficiaries of her commitments. Furthermore, she will be painfully aware that she will never feature in the intended audience of Bob's commitments, and so will cease to trust him in return.

Secondly, policing language means that people can't signal their trustworthiness as they would choose, making accurate placing of trust harder. Furthermore, the identity theorist does not attempt to interpret signals in good faith. By putting the worst construction on any language that he has not vetted, he opens up a gulf, and demands that it be bridged only on his own terms.

Thirdly, rejecting social standards and signalling systems for trustworthiness and replacing them with local standards of particular groups, enabling them to "express themselves authentically," is yet another recipe for disconnection rather than connection. For example, the ParliREACH workplace equality network for those working in the British Parliament reported that many ethnic minority staff were not comfortable "being themselves"' at work. "Some staff responded to the visible lack of diversity [in the Parliamentary estate] by feeling a need to conform and change their personas at work to minimise aspects of their racial or ethnic backgrounds. In one group colleagues spoke about this as a 'need to be invisible' resulting in a 'death to self' while at work" (ParliREACH 2019, 2). The report argued for categorising staff in racial terms, and enabling them to "be themselves."

The problem here is that being oneself is not conducive to a working atmosphere. No-one is herself at work, as Erving Goffman chronicled many years ago. A team is bound by "dramaturgical cooperation" – we find ourselves in roles which we must play to be on the team (Goffman 1959), and which communicate and even define trustworthiness in that context. Workplace roles no doubt could be more or less inclusive, and in many places may be unnecessarily prescriptive, but ultimately one cannot "be oneself" at work. The point of a workplace is to get something done with limited human and other resources, not self-expression. As David Lammy once pointed out, "as in any walk of life, the Westminster dress code is also about fitting in" (Lammy 2019). Indeed, and therefore helpful to those who follow it.

Lammy, however, thought it a bad thing. He took part in the 56 Black Men project,[4] in which prominent black citizens were photographed wearing hoodies, to counter the problem that "distrust of black men in hoodies is endemic in the UK," as they are "seen as surly or intimidating" (Lammy 2019). This is probably true, but not the whole story. Young men (and women) wear hoodies to "fit in" with their peer group; they are simultaneously signals of *trustworthiness* toward the in-group, and of *rebelliousness* toward outsiders, excluding them from the audience of beneficiaries. The men who agreed to be photographed in the project were all eminently respectable, but all but a couple glared intimidatingly at the viewer, looking pretty terrifying themselves. It is not a secret that the bourgeoisie are terrified of young men in gangsta gear, which is why it is worn. It is fun to terrify, which is fine until you need a bank loan or go for a job interview.

Lammy himself had recognised the dangers of aggressive and negative signalling in the years before critical identity theory went mainstream.

> I still don't believe that every young boy that listens to [controversial British rapper] Giggs immediately wants to pick up a knife. "Edgy" music has always formed the cornerstone to any teenage rebellion. Most indulge in it precisely because adults like me don't like them doing so.
>
> But when I hear those lyrics blaring out of the mobile phones of school kids from the back of the 41 bus, I can't help but feel uncomfortable. The diet of hate, misogyny and violence feels too close to the bone in Tottenham. ... The music reinforces the idea that life is rigged against you, that you'll never succeed through hard work so you shouldn't bother. (Lammy 2011)

The earlier Lammy wrote more sense about an essentially similar issue. It is not helpful to young people to encourage them to express their disdain of wider society (not least because they will anyway). It is not enough to be trustworthy. Trustworthiness in its social context involves reaching out to one's audience, understanding them and making commitments that will help them, in a language or with signals that they will understand. Withdrawing into one's own *argot* will result in disconnection and lost opportunity (on both sides) for collaboration and cooperation. Trustworthiness is other-directed, and can't be squared, as Nietzsche pointed out, with the imperative to value your integrity over everything else.

The left has long disparaged Margaret Thatcher's statement that "there is no such thing as society." Yet those who assert their identity as they define it, however provocative, are the ones who really act as if there were no such thing as society. By demanding respect unconditionally, they refuse to accept that society, with its norms, taboos and mores, has any legitimate influence.

---

4 https://www.56blackmen.com/.

What used to be called "sympathetic social pressure" is now more likely to be labelled "microaggression." By encouraging idiosyncratic and aggressive self-assertion, critical identity theorists themselves contribute to the marginalisation of oppressed groups. They out-Thatcher Thatcher.

The fourth effect of identity politics is to give momentum to trustless trust. Oppression will supposedly disappear because trustless trust will leave no space for it in public life. In the ideal world, no-one will have the ability to be symbolically violent, because the language will not be there to enable it; the whiff of Orwell's Newspeak is in the politically correct air. Everyone will be trustworthy, because no-one will be trusted to use their discretion.

Such is the damage to trust caused by identity politics. It is a harmful virus infecting the body politic that ideally should be suppressed – but not by a culture warrior. The culture wars happen because mainstream politicians have not proved themselves capable of standing up to the identity theorists and the Puritans, and indeed many in the mainstream have Puritan sympathies. The result is that the more forthright stance of the Peasants is seen as more trustworthy with respect to the identity virus.

My point is not that there is no illegitimate, damaging or hurtful prejudice, nor that Western democracies are always fair and have solved all their problems. Of course there is much injustice, but there is also justice, prosperity, civility, tolerance, amity and a general social *milieu* that is preferable to virtually everywhere else, such that the West has become a magnet for those from other parts of the world who would like to improve their lives. On the other hand, identity politics directly assumes Rousseau's thesis that all societies above a minimal level of complexity are corrupt and corrupting. This is destructive of trust, but our inquiry has shown that trust is the *sine qua non* for the construction and support of any kind of community, even – or especially – those which are cosmopolitan and diverse. Our common humanity, celebrated by cosmopolitans such as Appiah (2018) and Mounk (2022), is a far more noble and attractive ideal than the identity jungle, but between these two extremes we must somehow erect the social infrastructure of trust.

To place trust is *to know the trustee and to choose her over others*. For trustworthiness to be a successful strategy, *the trustee must be known and chosen*. Inclusion and exclusion are two sides of the same socially essential coin.

13. Conclusion: the trust dividend

Suppose the various ideologies of Part III are fought off and trust is being restored in and across the political class. What should we do with it? What political project will it enable? The answer I suggest in this chapter is the reconfiguration of the state. To do that, there must be some consensus on what it must do and where its priorities must lie.

The inquiry must take the form of the construction of a national project within each democracy, identifying those pressures from within and dangers from without which loom across more than one electoral cycle. Such a project will aim to (a) secure the state and optimise its capacity to resist those pressures, (b) protect the private sphere, and (c) preserve the foundations of trust and trustworthiness, in order to render the project sustainable. It will conserve what I called in an earlier work the invariants of successful democratic societies (O'Hara 2011, 106–160).

I have argued that trust and conservatism are natural allies – high-trust communities will tend to foster their precious resources, while conservatives tend to conserve those institutions which help connect trust and trustworthiness. Conservatism is also a useful lens through which to view our problem. Let's pose Disraeli's famous "awkward" question: "What will you conserve?"

To be conserved are the Western liberal capitalist democracies that are under threat from outsiders and insiders, especially the Peasants and Puritans. But to do this, these democracies, sometimes labelled "the West," need to be conceived *not* as exemplifications of universal principles, but as *civilisations and traditions*, afloat in a revived world of civilisational states (Coker 2019), and alert to the combative geopolitics this exposes them to (Huntington 1996).

The means is not simply to double down on liberalism. As we saw in Chapter 10, liberalism as the universal principle of emancipation is corrosive of trust, opening the door to the Peasants and Puritans after ceding the firepower to drive them away. Neither is the answer to become post-liberal. What we need to do is defend liberal society with non-liberal (i.e. conservative) means, protecting the trust upon which it is founded.

Opposing parties need to trust each other (as competitors, not enemies), they need to trust their voters, and their voters need to trust them. In order to cooperate internationally, governments need to trust each other too. Trust here is less an issue of corruption (the West is far less corrupt than most places), than in the competence, realism and goodwill of all sides. It involves rediscovering the art of opposition (Ionescu & de Madariaga 1972), and steering the conduct of politics away from the rational but unproductive strategy of blame avoidance (Hood 2011). It involves ceding to political actors the discretion they need to address problems both in the short term (e.g. raising taxes against a manifesto promise to deal with an unanticipated issue) and the long (e.g. shifting the burden of public spending away from some functions and toward others).

Binding them with trustless trust cannot be the solution to a lack of trust, since under such circumstances no-one has the ability to demonstrate, or even signal, trustworthiness, while decision-making is shackled by the lack of discretion afforded.

THE PROBLEM

Western democracies face serious problems requiring strategic (and sometimes coordinated) responses. The international rules-based order, an architecture of their own design, is crumbling, partly because outsiders who benefited from it would now prefer a more transactional world in which they have more influence, and partly because post-Cold War America has failed to understand where its interests lie. The rise of counterweights such as Brazil, China, India, Russia, Saudi Arabia, South Africa, Turkey and the United Arab Emirates is unsurprising, and welcome, but destabilising nonetheless.

The Problems

This is not a definitive list of problems that Western governments have to face, but they seem especially salient to me. The problems here (a) must be tackled consistently in a timescale that transcends the electoral cycle, (b) require mutual trust within the political class and between political actors and citizens, (c) are necessary to shore up the foundations of social trust, and (d) require the authority, power and legitimacy of the state. These are not the only problems that democratic states will have to deal with, but they have existential import and a list like this one should be an agreed priority for democratic political actors.

Note that this is *not* about finding a centre ground or triangulating between parties. The task is to find *common* ground between parties on what is valuable in a nation and the state that regulates and administers it.

Democracies now have serious and genuine enemies, especially Russia on the Eastern flank of Europe, Iran in the Middle East, and China, which threatens not only Taiwan but all countries with interests in the South China Sea. The number of nuclear-armed states is growing, and deterrence is more complicated than it was when there were only two players. Given the geopolitical realities, defence spending in the West, including NATO, needs to be raised drastically across the board, including high-tech weapons, top-notch cybersecurity and more personnel (the NATO spending target of 2% of GDP really won't cut it, but the point anyway is to agree on defence capabilities, not to set an arbitrary target). AI-powered disinformation is a serious threat to trust, and needs to be countered, as does state-sponsored cybercrime, which grows ever-more sophisticated. Organised crime itself is a threat to the state. And, however the Ukraine war ends, Ukraine will have to be rebuilt (and Russia will not do it, even in territory it continues to hold after a ceasefire).

The climate is changing, and economies have to adapt, both to decarbonise, and to adjust public health and research and development priorities. The investments required for renewable energy sources and storage, nuclear power, and carbon capture and storage, will be large. A whole electric vehicle charging network needs to be built virtually from scratch in short order. There will be enormous shifts in property prices, as the burden of items such as flood insurance and efficiency upgrades moves from government subsidies to private markets.

Ukraine and the climate both highlight the importance of highly complex infrastructure for the conduct of modern life in digital modernity, ranging from power generation to the Internet to transport systems to waste management. Building, upgrading and running such systems has become a giant job, dependent on all sorts of physical and digital processes. The Russians have taught us in Ukraine (and earlier in Syria) that smashing these intricate arrangements is debilitating, expensive – and easy for a ruthless foe. The state has to defend infrastructure, but its development and maintenance is now too big a job for the state, and not profitable enough for the private sector, so means of cooperation, finance, procurement, operation, evaluation and future-proofing require planning, unpicking collective action problems and defusing conflicts of interest (Gil & Fu 2022).

Societies are aging, which means adapting health systems to long-term care and management of chronic conditions. Pensions need to adapt to the new reality of fewer workers funding more beneficiaries in a time of lower growth, and finance needs to be sustainable. Pandemic planning needs to be solid and lessons learned from COVID. Economics needs to get to grips with a world in which the proportion of workers in their productive primes is falling.

Asylum and immigration set complex problems. The Western democracies are virtually the only well-run, safe and tolerant places on Earth. The low standard of governance elsewhere is only increasing the flow of people, whether fleeing for their lives, looking to experience freedom, escaping endemic gang violence, or simply seeking an economy that can absorb and pay for their labour. Migration to the West has increased several-fold between the 2010s and the mid-2020s. This helps economic growth, relieves labour shortages and alleviates some of the issues of aging, but has a number of negative consequences too. It has little support from voters and citizens, threatening political trust and legitimacy. Organised crime is growing on the back of the movement of people. There seem to be few options between ceding control of borders and retaining security at inhumane levels. More diverse populations, with many recent arrivals, will find trust in shorter supply.

Europe, and Britain in particular, have problems of their own. They have depended for decades on American protection, but its dysfunctional politics means that it can no longer be relied upon. Europe may need to defend itself against nuclear threats, conventional invasion and grey-zone actions (hostile action short of war, such as sabotage). Britain, having utterly botched Brexit, needs to re-establish a *modus operandi* with the EU, which itself needs to get over its pique. Britain's health service, police service and criminal justice system, basic functions of the state, have been allowed to decay to an appalling extent.

Trust and Stability

These are all serious problems that only the state can address. It needs therefore to be laser-focused on them. Any solutions can only be partial, ongoing and subject to correction, meaning that policy horizons extend beyond the four or five year electoral cycle. Some are multi-generational. They must be absorbed into a robust and sustainable political culture that can outlive immediate preferences, transient majorities, media panics and changing bones of contention.

That gives us a single imperative, if the democracies are to retain their positive status and qualities. The political classes, insurgents and elites, need to swing behind agreed parameters for policies on these issues. That doesn't mean policy agreement, but does mean creating a consensus on what serves the interests of each nation. Such consensus, where it is achieved, also needs to be *maintained*, which means it can't merely be technocratic – it needs to be accepted by voters, and to speak to their concerns about the quality of their life and the story of their nation (King & Crewe 2013, 243–253).

How will greater and more comprehensive political trust help with addressing these problems? Broadly speaking, the commitments made by political actors to each other, and to citizens, and by Burkean representatives need to be long-term and respected. How lack of trust doomed Brexit was discussed in the Introduction, but trust breakdowns have many forms. Another type of failure is where political elites cooperate very well but forget to engage with voters, leading to car crashes like the Millennium Dome (King & Crewe 2013, 111–125) or ID cards (King & Crewe 2013, 223–239). Relatedly, there are projects that are deemed necessary by elites who fail to have the necessary frank and open conversations with directly affected groups. Such projects also tend to be nixed when short-term electoral considerations are unfavourable; examples include Heathrow's second runway, and the entire nuclear power industry in the UK. A third type is where a flagship project is developed without consensus on its value. Such a project is likely to be cancelled, as was the British high speed rail network HS2 (Johnson et al. 2024, 79), as predicted a decade in advance (King & Crewe 2013, 414). Trust needs to placed in long-term projects, for instance, by investments being made on the assumption of its future delivery. Value judgments need to be trustworthy – virtually every major public investment comes in late and way over budget, because governments encourage applicants to bid low to gain the contract. The present government can trumpet the great deal it got for taxpayers, leaving future governments with the problem of deciding how to pay for the inevitable overruns.

This is not simply a plea for "joined up government," where governmental silos are dismantled to address wide-ranging problems, nor "mission-led government," where bold challenges are set by the state, and involve researchers and businesses across the economy in their pursuit (Mazzucato 2020). Neither of those is built to survive a change of government (or even a reshuffle of ministers). Buy-in to the broad project – not the detail – needs to be across political divides and into the citizenry. That requires the broad trust that underpins honest and frank conversation. It needs to be robust against dissent for dissent's sake, and the nihilism of critical "progressivism." The decadence of ideological fragmentation, which cripples politicians' ability to perform even routine tasks, needs to be disciplined – all of us need to recognise the good faith of our opponents, and to demonstrate our own. We need "elaborate agreements on how to disagree" (Ionescu & de Madariaga 1972, 187).

We should also accept that long-term political stability needs to be shaken up every now and then to reduce the risk of sclerosis. The reset of Germany and Japan after WWII (Olson 1982, 75–117), the US after Vietnam and Watergate, and less drastically, the Thatcherite shake-up in Britain (O'Hara 2013) were important in setting those countries on a more productive course, painful and controversial though the changes were at the time. The creative destruction of capitalism also helps prevent extreme stasis and complacency (Schumpeter 1950). Such shocks cannot be planned, but have their value. But for them to succeed, trust and trustworthiness need to be restored quickly and constructively afterwards.

Demands on Politics

Addressing these problems does not assume a common good or purpose for the state – in Oakeshott's terms, to make of the state an enterprise association. Rather, they – or some other similar list – are basic issues that need to be resolved between parties to create and maintain a civil association in the conditions of digital modernity in which freedom for citizens is maximised and everyone can pursue their own idea of the good. So complex is 21st-century infrastructure and geopolitics that securing the state has become a colossal task in itself. It can no longer be a matter of a small state standing back to let civil society flourish.

Most of these problems will require investment, which means that, if the state is to address them effectively, it must stop spending on unnecessary things, or things it does badly. Increasing the size of the army is more important than ensuring that swimming pool attendants are diverse enough. There must be a rough cross-party settlement as to how much of the economy the state can gobble up. It must face up to the choice of taxing (reducing income, suppressing private investment), borrowing (raising interest rates and payments), printing money (raising inflation) or cutting superfluities (undermining services and creating a loud constituency of losers). Stability, important both for trust and for a successful investment climate, requires agreement about maximum and minimum levels of government intervention. If sacrifices need to be made, they should be argued for honestly and agreed across political divides, to reduce the chances of being undercut by populists.

It is not an easy task. It cannot be achieved by the Peasants or the Puritans, more concerned with kicking lumps out of each other and burying trust. Partisanship is the enemy within, ideology unrestrained is destructive of trust. Both Peasant and Puritan seem to think that the foundational substrates of free, stable democratic societies will survive their own antics. This is the worst kind of decadence: self-indulgent, complacent, and destructive of the societies whose protection they have benefited from and whose products they have enjoyed.

TOWARDS A SOLUTION

What to do is not trivial, and in one sense is not in anyone's gift. It needs social change, which is a problem. As Hume put it, "all plans of government, which suppose great reformation in the manners of mankind, are plainly imaginary" (Hume 1987b, 514). However, the excessive pessimism of a recent book review (of Kearney 2023) by *The Economist*'s Free Exchange columnist deserves to be debunked (Free Exchange 2023). The reviewer was impressed, but "For all the striking statistics in the book, the author does not offer an especially rousing call to arms." Why not? Because she demanded norm changes and "it is doubtful that governments can do all that much to alter social norms, just as they hold little responsibility for the fact that they changed in earlier decades," while "knowledge of what produces successful adults is most useful to individuals planning their own lives."

This is an extraordinary argument. These statements are of course true, but incredibly the reviewer appears to believe that between state and citizen nothing affects norms at all; if people don't spontaneously shift *en masse*, and if government coercion is powerless, all options are exhausted. But there are plenty of institutions, associations and little platoons in the private sphere between state and citizen that are immensely influential. If these are allowed to flourish, not whittled away by attacks on freedom of association and incursions of the public sector, then they may influence norms in a positive direction.

More to the point, what about the enormous influence on social norms of teachers, college educators, social researchers and journalists (including the *Economist* reviewer)? If they are impressed or persuaded by a book, and trumpet its virtues and cease to promote policies it discredits, wouldn't that help? If they were to do it in sufficient numbers, in schools, colleges, the media and social media, it might actually help to shift the norm in the direction urged by the evidence. That's how an intellectual community is supposed to work within civil society (Mounk 2022, 249).

It doesn't guarantee that norms will shift in the right direction, but it can only help. The liberal elite has often acted as a vanguard for social attitudes more widely, sometimes to the detriment of society as a whole, as long-cherished institutions, traditions and behaviours have been sidelined by new liberties and a general air of licence. If commentators were prepared to boost rather than undermine trust, celebrate and defend their nations instead of critique them, and make the good will of political opponents the default assumption, it may be that such attitudes, filtered through the media and social media, could diffuse through the population, undoing some of the damage of the culture wars.

Countering the Peasants and the Puritans

In the Introduction, the Peasants and Puritans were characterised in terms of five ideas which need to be countered by trust-based politics. The first was a negative attitude toward the West – the declinism of the Peasants, and the rejectionism of the Puritans. The political and intellectual elites of most Western democracies are ashamed of these most successful of societies, and their guilt is propagated through the universities and arts. "Faced with history's rich parade of horrors, many people understandably despair for their country, their culture, their race." Understandably? "Most countries really do have shameful chapters in their histories" (Mounk 2022, 285). Sure. But is this the only perspective (cf. McCloskey 2010)?

The path dependencies of these shameful chapters make it difficult to disentangle the extent to which they contributed to the successes of those societies in the present. It is only in the West that this shame is very strong, although it is not the only place in the world with a ruthless colonial past. The Turks reject charges of genocide against the Armenians, the Japanese barely accept responsibility for their wartime actions against China, Korea and elsewhere, and the Russians happily gloss over all sorts of atrocities in their bloody history, up to the present day. There must surely be a middle way between ignoring the past and self-mortification on its account.

Mounk accepted that "Compared with the long sweep of human history, what is striking about the current state of France or Japan, Australia or the United States is not so much the things these societies continue to get wrong …; it is what they get right. … A fair assessment of these countries' current condition needs to feature the good alongside the bad" (Mounk 2022, 286). OK, but what counts as good and bad? In the bad he cited racist violence, and we can certainly agree that that is a stain on a community. The only example of good behaviour he included was atoning for racist history. In other words, he conceded to the critical theorists that the only way to evaluate current behaviour is with reference to historical evils.

But good behaviour goes way beyond anti-racism. In a good society, people pay taxes, debts and bills, don't commit many crimes, aren't very violent, resolve disputes lawfully, avoid conflict, work hard and productively, make decades-long plans for pensions, mortgages, educating children and caring for parents, and generally make the nation function. That is the small-scale element of a national vision, based on citizens' behaviour and deportment. It may not be edgy, but it is the bourgeois lifestyle that sustains nations – trusting, trustworthy, and cementing trust and trustworthiness.

The second idea upon which Peasant and Puritan agree is the state of emergency (although they don't agree about what constitutes the emergency). The corrective to this is to understand and manage risks, eschew the decadent pleasures of ideological conflict and get down to the hard work of normal politics. "The posttotalitarian democracies therefore need to recover the ethical force of the dignity of the person and of reasoned participation in political life in order to shield themselves from the temptations of an intolerant communitarianism, but also to redeem the public sphere from the quasi-tyrannical sway of a possessive and politically apathetic individualism" (Urbinati 2015, 25). We need to mobilise the trust relationships between persons, with their interlaced, reciprocally connected network of commitments. These flourish outside the glare of scrutiny, so one route to protecting our liberal societies is to promote and protect the private sphere.

The third aspect, Manichaeanism, should be rejected outright. Schmitt's philosophy of friend and enemy is inimical to trust. This does not mean that politics should become technocratic (i.e. depoliticised); the trick is managing opposition (and defeat) with rationality and humanity (Ionescu & de Madariaga 1972).

In their fourth characteristics, both the Peasants and Puritans wish to replace current elites, the former with a charismatic leader, and the latter with a vanguard cultivated in public institutions. Here, they go rather more with the grain of current society, at all levels of which the habit of rejection is taken as a marker of independence of mind (which it clearly is not).

Widespread rejectionism is not promising. As noted in the Introduction, cynicism about politics has always been with us. If we try to give it a more positive spin, however, it may be interpreted as a rejection of beggar-thy-neighbour politics in particular, a disavowal of unseemly conflict between elite ideological cliques, and a preference for civility. Naturally, this precludes those types of politics that are destructive of trust, such as far right populism, critical "progressivism" and identity politics. Self-styled activists prefer to play beggar-thy-neighbour.

> For these activists, "being political" is in part a defiant stance against contemporary political culture. "Political" citizens reject the dominant discourse that appropriate engagement is polite and nonconfrontational and that politics should benefit all people equally. (Bennett et al. 2013, 542)

Defenders of liberal societies are too often prepared to cede that dominant discourse in the face of protest. For instance, Mounk asserted that smaller groups will assume that a government is illegitimate if they feel disrespected, even warning of civil war (2022, 104). This feels like a capitulation to the politics of inclusion, and a neglect of the exclusivity that is often a priority for a majority of people in successful societies (2022, 287). Does every incomer, or wannabe incomer, get an immediate say?

Abandoning civility leaves a vacuum that cedes politics to the uncivil. By now it is a stretch to call a contemporary political culture that is polite, civilised, and sustaining of community and trust the "dominant discourse," because the Peasants and Puritans have undermined it and expelled it from the public sphere. But rediscovery of our habits of politeness will help foster the virtues of trust and trustworthiness and cultivate the relations and discourses we value.

The fifth characteristic of both Peasantry and Puritanism is the Gramscian metapolitics of cultural hegemony. Gramsci himself grumbled that liberal intellectuals "extend the world view of the rulers to the ruled, and thereby secure the 'free' consent of the masses to the law and order of the land" (Bates 1975, 353), and both Peasants and Puritans hope to do the same thing. Rather, the process should be bottom-up. The bourgeoisie and working class – not only intellectuals, but professionals, businesspeople, families, workers, savers, the law-abiding – whose lives and livelihoods depend on cooperation, collaboration, privacy, respect and work, should ideally be able to extend their own practices of trust, trustworthiness and tolerance from their own private interactions back into the public sphere whence they have been exiled.

The End of Trustless Trust

For many liberal thinkers, trust threatens autonomy. For example, trust in government is apparently widespread in China, leading to great stability but not very much independence of mind (Urbinati 2015, 181). A fair point, but hardly decisive in a Western context. The West's Christian tradition eventually spawned liberal individualism (Siedentop 2014; Spencer 2016), while the Asians are influenced by their Confucian past. In the Confucian tradition, emphasis is placed on citizens signalling their trustworthiness to authority, but we can ignore the non-existent possibility of that catching on in the West. My call is rather for there to be less *mistrust* as a knee-jerk reaction in the West, and greater sensitivity from journalists and academics (and voters) to its positive aspects.

Urbinati wrote that "political action in democracy has an ineradicably voluntary character that cannot be assimilated by a virtue transmuted into habit and routine" (2015, 151). She was arguing correctly that citizenship cannot be disinterested, but her specific point is too strong. We saw earlier that trust and trustworthiness tap into virtuous qualities, are supportive of communities, and are sustained precisely by transmutation into habit. There is a major difference between trustless trust that removes discretion, and the regularities and expectations that develop alongside routine, habit and convention. One can always diverge from one's habits, but where they implement virtuous behaviour, they are usually robust (except when challenged by negative stimuli, such as trust misplaced or trustworthiness denied).

While they have used their independent thought to concoct their worldview ("dare to know!"), Puritans and their fellow travellers on the left are keen not to extend the privilege more widely. Their low trust of citizens is realised by trustless trust, implemented by different mechanisms, including nudging, engineering and digitising the environment, injecting ideology into early education, and haranguing and shaming, to moderate the citizenry's untutored instincts. The democratic state's precepts are increasingly disconnected from the common sense of ordinary folk (Canovan 2004; Hertogh 2018), creating space for populists. Joseph Conrad remarked that the "temporal world rests on a few very simple ideas; so simple that they must be as old as the hills" (Conrad 1912, xxi), and this moral clarity or legibility is important for most of us. The Puritans, on the other hand, tend to complicate or even undermine such norms, sympathising with the drug-addled robber, rather than with the hard-working shop owner injured attempting to preserve her property. What are perceived as perfectly reasonable actions are penalised, and what seem like normal attitudes and feelings are disparaged by official messaging that appears unpatriotic, antisocial or disrespectful.

The populist wave shows that there are limits to how far circumscribing citizens' and voters' options with trustless trust mechanisms can go. The Peasants have been successful electorally by ignoring the decencies of due process and focusing on common sense goals such as security and patriotism. If we are to steer a path between the illiberalism of the Peasants and the *hauteur* of the Puritans, the political class needs to be more sensitive to public opinion, providing leadership that is empathetic, collaborative and focused on the long term. But this requires trust that the opposition will share and respect agreed long-term goals.

With the Grain

Mounk mused upon the likely future trajectory of identity politics (2023, 265–286), and is probably correct to say that it will neither triumph nor be vanquished, but there will be a long struggle against it. In such a world, it is important that we distinguish between good faith and bad faith, trusting and mistrusting politics. Conflict may be inevitable if we are to reduce the Puritans' influence.

The Peasants tend towards incompetence, internecine rivalries and impracticability, and they often wither away as they approach power. When they do not, they are buoyed by the same forces: the influence of the Puritans, and a political elite that imperiously disregards popular opinions. It is quite possible that some of the Peasants might be co-opted into the defence of the vernacular Western tradition.

The elite must reconfigure itself to defend liberal democracy. This will involve facing down the illiberal Puritans, and reacquiring respect for civil political culture. This latter task will involve constructively helping craft a political vision that will transcend party and survive electoral changes, as argued at the beginning of this chapter. This will be a matter of trusting and listening to citizens rather than dictating to or nudging them (King & Crewe 2013, 243–253). It is the citizens who make up a community.

Perhaps loud naysayers will continue to attract attention via social media, disinformation and so on. Perhaps the reasonable political class will wither away from the combined force of Peasant and Puritan, as it appears to be doing in the United States. Maybe, but the first stage of the recovery of the democracies is the realisation that these are the most successful societies on the planet, and through history. The second needs to be the rediscovery of trust, and an understanding of how that requires the *fostering* of community, not its reconfiguration. The third stage is to use those insights to look unflinchingly and see where the threats to communities lie – geopolitical, demographic, meteorological, and so on – without piety or unrealistic idealism.

Liberalism mythologises itself as a provider of universal goods, yet all over the world democracy is accepted to the extent that it is seen to facilitate stronger, more prosperous, more resilient, more welcoming and pleasanter communities. It follows that if the abstract goods are taken for granted, and practical effects are ignored or glossed over, democracy will become unpopular, as sometimes happens (Foweraker 2016; Graham & Svolik 2020; Kilavuz & Sumaktoyo 2020; Lu & Chu 2021; Fossati et al. 2022). The depth of democratic tradition and history are unsurprisingly important (Chlouba et al. 2022).

If liberals can compromise and understand that their ideals depend on more fundamental structures, we can defend liberal society against the Peasants and the Puritans beyond the electoral cycle with a coalition of liberals and non-liberals alike. Sometimes liberal parties and policies will be in the ascendant, sometimes conservative ones. Criticism and debate will never cease. But the conservation of the basic relations of trust and trustworthiness that underpin these highly successful communities should be the priority.

References

Uriel Abulof (2017). "Introduction: why we need Maslow in the twenty-first century," *Society*, 54(6), 508–509, https://doi.org/10.1007/s12115-017-0198-6.

Jon Agar (2024). "Science," in Anthony Seldon & Tom Egerton (eds.), *The Conservative Effect 2010–2024: 14 Wasted Years?* Cambridge: Cambridge University Press, 326–353.

Neil M. Agnew, Kenneth M. Ford & Patrick J. Hayes (1997). "Expertise in context: personally constructed, socially selected and reality-relevant?" in Paul J. Feltovich, Kenneth M. Ford & Robert R. Hoffman (eds.), *Expertise in Context*, Menlo Park CA: American Association for Artificial Intelligence, 219–244.

Sara Ahmed (2007). "A phenomenology of whiteness," *Feminist Theory*, 8(2), 149–168, https://doi.org/10.1177/1464700107078139.

Shankkar Aiyar (2017). *Aadhaar: A Biometric History of India's 12-Digit Revolution*, Chennai: Westland.

Pepijn Al (2023). "(E)-trust and its function: why we shouldn't apply trust and trustworthiness to human-AI relations," *Journal of Applied Philosophy*, 40(1), 95–108, https://doi.org/10.1111/japp.12613.

Muhsin Ali, Karim Khan & Muhammad Saeed Meo (2022). "Trust in post-conflict life: evidence from a conflict zone in Pakistan," *Journal of Public Affairs*, 22(2), e2403, https://doi.org/10.1002/pa.2403.

Darcy W.E. Allen, Chris Berg & Mikayla Novak (2018). "Blockchain: an entangled political economy approach," *Journal of Public Finance and Public Choice*, 33(2), 105–125, https://doi.org/10.1332/251569118X15282111163993.

Gordon W. Allport (1954). *The Nature of Prejudice*, Boston: Addison-Wesley.

Sam Altman (2021). *Moore's Law for Everything*, https://moores.samaltman.com/.

Rashid Ameer & Radiah Othman (2022). "Gender, fraud opportunity, and rationalisation," *Journal of Criminology*, 55(1), 81–105, https://doi.org/10.1177/26338076211065180.

Benedict Anderson (1991). *Imagined Communities: Reflections on the Origin and Spread of Nationalism*, London: Verso.

Octavio Ansaldi & María Pardo-Vergara (2020). "What constitution? On Chile's constitutional awakening," *Law and Critique*, 31(1), 7–39, https://doi.org/10.1007/s10978–020–09260–0.

G.E.M. Anscombe (1958). "Modern moral philosophy," *Philosophy*, 33(124), 1–19, https://doi.org/10.1017/S0031819100037943.

Andreas M. Antonopoulos (2015). *Mastering Bitcoin: Unlocking Digital Cryptocurrencies*, Sebastopol CA: O'Reilly Media.

Marco Antonsich (2010). "Searching for belonging – an analytical framework," *Geography Compass*, 4(6), 644–659, https://doi.org/10.1111/j.1749–8198.2009.00317.x.

Fredy Andres Aponte-Novoa, Ana Lucila Sandoval Orozco, Ricardo Villanueva-Polanco & Pedro Wightman (2021). "The 51% attack on blockchains: a mining behavior study," *IEEE Access*, 9, 140549–140564, https://doi.org/10.1109/ACCESS.2021.3119291.

Kwame Anthony Appiah (2005). *The Ethics of Identity*, Princeton: Princeton University Press.

Kwame Anthony Appiah (2018). *The Lies That Bind: Rethinking Identity: Creed, Country, Colour, Class, Culture*, Profile: London.

Anne Applebaum (2024). *Autocracy, Inc.: The Dictators Who Want to Run the World*, New York: Allen Lane.

Benjamin Arditi (2003). "Populism, or, politics at the edges of democracy," *Contemporary Politics*, 9(1), 7–31, https://doi.org/10.1080/1356977032000072486.

Aristotle (2011). *Aristotle's Nicomachean Ethics*, Chicago: University of Chicago Press.

John Armstrong (2024). "The QAA's subject benchmarks and critical pedagogy: the example of 'gateway to King's'," *British Educational Research Journal*, https://doi.org/10.1002/berj.4059.

Olov Aronson (2021). "Victimhood in Swedish political discourse," *Discourse and Society*, 32(3), 292–306, https://doi.org/10.1177/0957926520977216.

Kenneth Arrow (1974). *The Limits of Organization*, New York: Norton.

Joaquín Artés (2013). "Do Spanish politicians keep their promises?" *Party Politics*, 19(1), 143–158, https://doi.org/10.1177/1354068811407581.

Jan Assmann (2004). "Confession in Ancient Egypt," in Adriana Destro & Mauro Pesce (eds.), *Ritual and Ethics: Patterns of Repentance: Judaism, Christianity, Islam: Second International Conference of Mediterraneum*, Paris: Peeters, 1–12.

Jan Assmann (2006a). "Introduction: what is 'cultural memory'?" in *Religion and Cultural Memory*, Stanford: Stanford University Press, 1–30.

Jan Assmann (2006b). "Remembering in order to belong: writing, memory and identity," in *Religion and Cultural Memory*, Stanford: Stanford University Press, 81–100.

Fatma Aydemir (2024). "In Germany, football has made nationalism cool again. That's why I'm dreading the Euros," *The Guardian*, 13th June, 2024, https://www.theguardian.com/commentisfree/article/2024/jun/13/in-germany-football-has-made-nationalism-cool-again-thats-why-im-dreading-the-euros.

Arshdeep Bahga & Vijay K. Madisetti (2016). "Blockchain platform for Industrial Internet of Things," *Journal of Software Engineering and Applications*, 9(10), 533–546, http://dx.doi.org/10.4236/jsea.2016.910036.

Annette C. Baier (1994a). "What do women want in a moral theory?" in *Moral Prejudices: Essays on Ethics*, Cambridge MA: Harvard University Press, 1–17.

Annette C. Baier (1994b). "Unsafe loves," in *Moral Prejudices: Essays on Ethics*, Cambridge MA: Harvard University Press, 33–50.

Annette C. Baier (1994c). "Hume, the women's moral theorist?" in *Moral Prejudices: Essays on Ethics*, Cambridge MA: Harvard University Press, 51–75.

Annette C. Baier (1994d). "Hume, the reflective women's epistemologist?" in *Moral Prejudices: Essays on Ethics*, Cambridge MA: Harvard University Press, 76–94.

Annette C. Baier (1994e). "Trust and antitrust," in *Moral Prejudices: Essays on Ethics*, Cambridge MA: Harvard University Press, 95–129.

Annette C. Baier (1994f). "Trust and its vulnerabilities," in *Moral Prejudices: Essays on Ethics*, Cambridge MA: Harvard University Press, 130–151.

Annette C. Baier (1994g). "Sustaining trust," in *Moral Prejudices: Essays on Ethics*, Cambridge MA: Harvard University Press, 152–182.
Annette C. Baier (1994h). "Trusting people," in *Moral Prejudices: Essays on Ethics*, Cambridge MA: Harvard University Press, 183–202.
Annette C. Baier (1994i). "Violent demonstrations," in *Moral Prejudices: Essays on Ethics*, Cambridge MA: Harvard University Press, 203–223.
G.P. Baker & P.M.S. Hacker (1985). *Wittgenstein: Rules, Grammar and Necessity*, Oxford: Basil Blackwell.
Stephanie Alice Baker (2022). "Alt. health influencers: how wellness culture and web culture have been weaponised to promote conspiracy theories and far-right extremism during the COVID-19 pandemic," *European Journal of Cultural Studies*, 25(1), 3–24, https://doi.org/10.1177/13675494211062623.
Stephanie Alice Baker & Michael James Walsh (2022). "'A mother's intuition: it's real and we have to believe in it': how the maternal is used to promote vaccine refusal on Instagram," *Information, Communication and Society*, 26(8), 1675–1692, https://doi.org/10.1080/1369118X.2021.2021269.
Trevor Baker (2020). "Why you shouldn't book with Opodo," *Which?* 12th Oct, 2020, https://www.which.co.uk/news/article/why-you-shouldnt-book-with-opodo-arkV00e6l3eD.
Michael Barkun (2013). *A Culture of Conspiracy: Apocalyptic Visions in Contemporary America*, 2nd edition, Berkeley: University of California Press.
Barry Barnes & David Bloor (1982). "Relativism, rationalism and the sociology of knowledge," in Martin Hollis & Steven Lukes (eds.), *Rationality and Relativism*, Oxford: Basil Blackwell, 21–47.
Hannah Barnes (2023). *Time To Think: The Inside Story of the Collapse of the Tavistock's Gender Service for Children*, Swift Press.
Randy E. Barnett (2012). "Contract is not promise: contract is consent," *Suffolk University Law Review*, 45(3), 647–665.
Thomas R. Bates (1975). "Gramsci and the theory of hegemony," *Journal of the History of Ideas*, 36(2), 351–366, https://doi.org/10.2307/2708933.
BBC News (2015). "My life with a face tattoo," *Times Two*, 17th Oct, 2015, https://www.bbc.com/news/av/magazine-34544013.
Kim Beauchesne & Alessandra Santos (eds.) (2017). *Performing Utopias in the Contemporary Americas*, New York: Palgrave Macmillan.
Koa Beck (2021). *White Feminism: From The Suffragettes to Influencers and Who They Leave Behind*, New York: Atria.
Derrick A. Bell Jr. (1980). "*Brown v. Board of Education* and the interest-convergence dilemma," *Harvard Law Review*, 93(3), 518–533, https://doi.org/10.2307/1340546.
Derrick Bell (1993). "The racism is permanent thesis: courageous revelation or unconscious denial of racial genocide," *Capital University Law Review*, 22(3), 571–587.
Derrick Bell (2004). *Silent Covenants:* Brown v Board of Education *and the Unfulfilled Hopes for Racial Reform*, New York: Oxford University Press.
Walter Benjamin (2019). "Theses on the philosophy of history," in *Illuminations: Essays and Reflections*, Boston MA: Mariner Books, 196–209.
Elizabeth A. Bennett, Alissa Cordner, Peter Taylor Klein, Stephanie Savell & Gianpaolo Baiocchi (2013). "Disavowing politics: civic engagement in an era of political skepticism," *American Journal of Sociology*, 119(2), 518–548, https://doi.org/10.1086/674006.

Jeffrey M. Berry & Sarah Sobieraj (2014). *The Outrage Industry: Political Opinion Media and the New Incivility*, New York: Oxford University Press.

Bharat Bhushan, G. Sahoo & Amit Kumar Rai (2017). "Man-in-the-middle attack in wireless and computer networking — a review," in *3rd International Conference on Advances in Computing, Communication & Automation (ICACCA)*, https://doi.org/10.1109/ICACCAF.2017.8344724.

Cristina Bicchieri (2017). *Norms in the Wild: How to Diagnose, Measure, and Change Social Norms*, New York: OxfordUniversity Press.

Christopher Bickerton & Carlo Invernizzi Accetti (2017). "Populism and technocracy: opposites or complements?" *Critical Review of International Social and Political Philosophy*, 20(2), 186–206, https://doi.org/10.1080/13698230.2014.995504.

Andrew Blick & Peter Hennessy (2019). *Good Chaps No More? Safeguarding the Constitution in Stressful Times*, London: The Constitution Society, https://consoc.org.uk/wp-content/uploads/2019/11/FINAL-Blick-Hennessy-Good-Chaps-No-More.pdf.

Mark Blyth (2013). *Austerity: The History of a Dangerous Idea*, New York: Oxford University Press.

Anna Boncompagni (2022). *Wittgenstein on Forms of Life*, Cambridge: Cambridge University Press.

Matilda Boseley (2023). "ADHD has become an identity, not just a disorder. We need a new way to talk about it," *The Guardian*, 20th Sept, 2023, https://www.theguardian.com/society/2023/sep/21/adhd-has-become-an-identity-not-just-a-disorder-we-need-a-new-way-to-talk-about-it.

Matteo Botto & Lucas Gottzén (2024). "Swallowing and spitting out the red pill: young men, vulnerability, and radicalization pathways in the manosphere," *Journal of Gender Studies*, 33(5), 596–608, https://doi.org/10.1080/09589236.2023.2260318.

Pierre Bourdieu (1977). *Outline of a Theory of Practice*, Cambridge: Cambridge University Press.

Pierre Bourdieu (1991). *Language and Symbolic Power*, Cambridge: Polity Press.

Pierre Bourdieu (1998). *Practical Reason*, Cambridge: Polity Press.

Pierre Bourdieu (2018). "The forms of capital," in Mark Granovetter & Richard Swedberg (eds.), *The Sociology of Economic Life*, 3rd edition, Abingdon: Routledge, 78–92.

Pierre Bourdieu et al (1999). *The Weight of the World: Social Suffering in Contemporary Society*, Cambridge: Polity Press.

Julian Bourg (2017). *From Revolution to Ethics: May 1968 and Contemporary French Thought*, 2nd edition, Montreal: McGill-Queen's University Press.

Gwen Bouvier & David Machin (2021). "What gets lost in Twitter 'cancel culture' hashtags? Calling out racists reveals some limitations of social justice campaigns," *Discourse and Society*, 32(3), 307–327, https://doi.org/10.1177/0957926520977215.

Jean Braucher (2012). "The sacred and profane contracts machine: the complex morality of contract law in action," *Suffolk University Law Review*, 45(3), 667–693.

Robert Brenneman (2014). "Wrestling the Devil: conversion and exit from Central American gangs," *Latin American Research Review*, 49(S1), 112–128, https://doi.org/10.1353/lar.2014.0062.

Shannon Brincat & Michael Löwy (2023). "Che lives! The legacy of Che Guevara in world politics," *Globalizations*, 20(8), 1447–1463, https://doi.org/10.1080/14747731.2023.2213497.

Kristin A. Broussard & Helen C. Harton (2018). "Tattoo or taboo? Tattoo stigma and negative attitudes toward tattooed individuals," *Journal of Social Psychology*, 158(5), 521–540, https://doi.org/10.1080/00224545.2017.1373622.

Rogers Brubaker (2016). "The Dolezal affair: race, gender, and the micropolitics of identity," *Ethnic and Racial Studies*, 39(3), 414–448, https://doi.org/10.1080/01419870.2015.1084430.

Lori Bruce & Ruth Tallman (2021). "Promoting racial equity in COVID-19 resource allocation," *Journal of Medical Ethics*, 47(4), 208–212.

Heather Brunskell-Evans (2020). *Transgender Body Politics*, North Geelong: Spinifex.

F.H. Buckley (2018). *The Republican Workers Party: How the Trump Victory Drove Everyone Crazy, and Why it was Just What We Needed*, New York: Encounter Books.

Edmund Burke (1881). "Mr. Burke's Speech at the Guildhall, in Bristol," in Matthew Arnold (ed.), *Letters, Speeches and Tracts on Irish Affairs by Edmund Burke*, London: Macmillan & Co, 116–181.

Edmund Burke (1999a). "Reflections on the Revolution in France," in Francis Canavan (ed.), *Select Works of Edmund Burke: Volume 2*, Indianapolis: Liberty Fund, 85–365.

Edmund Burke (1999b). "Speech to the electors of Bristol," in Francis Canavan (ed.), *Select Works of Edmund Burke: Miscellaneous Writings*, Indianapolis: Liberty Fund, 3–13.

Andrew Bushnell (2022). *Order and the Reason to be Conservative*, PhD thesis, University of Melbourne, http://hdl.handle.net/11343/333970.

Laura Buss & Karen Hodges (2017). "Marked: tattoo as an expression of psyche," *Psychological Perspectives*, 60(1), https://doi.org/10.1080/00332925.2017.1282251.

Judith Butler (2012). ""Occupy as form," *ARC Muses*, blog, http://arcdirector.blogspot.com/2012/02/occupy-as-form-judith-butler.html.

Jesse L. Byock (2003). "Feuding in Viking age Iceland's great village," in Warren C. Brown & Piotr Górecki (eds.), *Conflict in Medieval Europe: Changing Perspectives on Society and Culture*, Aldershot: Ashgate, 229–241.

Margaret Canovan (2004). "Populism for political theorists?" *Journal of Political Ideologies*, 9(3), 241–252, https://doi.org/10.1080/1356931042000263500.

Cao Xueqin (1977). *The Story of the Stone Volume II*, London: Penguin.

Catherine Carr (2024). "From doomscrolling to sex: being a boy in 2024," *The Guardian*, 5th May, 2024, https://www.theguardian.com/society/article/2024/may/05/from-doomscrolling-to-sex-being-a-boy-in-2024.

Helena Carthew (2023). *House of Commons: Hours Sat and Late Sittings*, House of Commons Library Research Briefing SN02226, https://researchbriefings.files.parliament.uk/documents/SN02226/SN02226.pdf.

Hilary Cass (2024). *Independent Review of Gender Identity Services for Children and Young People: Final Report*, NHS England, https://cass.independent-review.uk/home/publications/final-report/.

Quassim Cassam (2019). *Conspiracy Theories*, Cambridge: Polity Press.

Alessandra Cassar, Pauline Grosjean & Sam Whitt (2013). "Legacies of violence: trust and market development," *Journal of Economic Growth*, 18(3), 285–318, https://doi.org/10.1007/s10887–013–9091–3.

Manuel Castells (2000). *The Information Age: Economy, Society and Culture Volume 1: The Rise of the Network Society*, 2nd edition, Malden: Blackwell.

Manuel Castells (2019). *Rupture: The Crisis of Liberal Democracy*, Cambridge: Polity Press.

Leyland Cecco (2021). "'We know who we are': Inuit row raises questions over identity and ancestry," *The Guardian*, 24th Oct, 2021, https://www.theguardian.com/world/2021/oct/24/canada-inuit-nunatukavut.

Leyland Cecco (2024). "Is a previously unheard-of First Nation just Canada's latest Pretendian case?" *The Guardian*, 9th May, 2024, https://www.theguardian.com/world/article/2024/may/09/canada-first-nation-kawartha-lakes-validity.

Chainalysis Team (2024). *Activity Down as Scamming and Stolen Funds Fall, But Ransomware and Darknet Markets See Growth*, Chainalysis, https://www.chainalysis.com/blog/2024-crypto-crime-report-introduction/.

Henry 'Chips' Channon (2021). *The Diaries 1918–38*, London: Hutchinson.

William G. Chase & Herbert A. Simon (1973). "Perception in chess," *Cognitive Psychology*, 4(1), 55–81, https://doi.org/10.1016/0010–0285(73)90004–2.

Vladimir Chlouba, Daniel S. Smith & Seamus Wagner (2022). "Early statehood and support for autocratic rule in Africa," *Comparative Political Studies*, 55(4), 688–724, https://doi.org/10.1177/00104140211036031.

Sumi Cho, Kimberlé Williams Crenshaw, and Leslie McCall (2013). "Toward a field of intersectionality studies: theory, applications, and praxis," *Signs*, 38(4), 785–810, https://doi.org/10.1086/669608.

Usman W. Chohan (2019). "Are cryptocurrencies truly trustless?" in Stéphane Goutte, Khaled Guesmi & Samir Saadi (eds.), *Cryptofinance and Mechanisms of Exchange: the Making of Virtual Currency*, Cham: Springer, 77–89, https://doi.org/10.1007/978-3-030–30738–7_5.

Clayton M. Christensen (1997). *The Innovator's Dilemma: When New Technologies Cause Great Firms to Fail*, Boston MA: Harvard Business News Press.

Amy Chua (2003). *World on Fire: How Exporting Free Market Democracy Breeds Ethnic Hatred and Global Instability*, London: Heinemann.

Cicero (1928). *De Republica*, London: Heinemann.

Harold D. Clarke, Matthew Goodwin & Paul Whiteley (2017). *Brexit: Why Britain Voted to Leave the European Union*, Cambridge: Cambridge University Press.

Steve Clarke (2017). "A prospect theory approach to understanding conservatism," *Philosophia*, 45(2), 551–568, https://doi.org/10.1007/s11406–017–9845–9.

Adam Cleary (2016). "Here is how the DAO will soon become the greatest threat banks have ever faced," *Frontera*, 25th May, 2016, https://fronteranews.com/news/dao-will-soon-become-greatest-threat-banks-ever-face/.

Shelley Cobb (2018). "'I'd like y'all to get a black friend': the politics of race in *Friends*," *Television and New Media*, 19(8), 708–723, https://doi.org/10.1177/1527476418778420.

Jennifer E. Cobbina-Dungy & Delores Jones-Brown (2023). "Too much policing: why calls are made to defund the police," *Punishment and Society*, 25(1), 3–20, https://doi.org/10.1177/14624745211045652.

Lorraine Code (1987). "Second persons," in *Canadian Journal of Philosophy Supplementary Volume*, 13, 357–382, https://doi.org/10.1080/00455091.1987.10715942.

Wilfred U. Codrington III (2022–23). "The phantasm of principle," *Kentucky Law Journal*, 111(4), 651–691.

Deborah A. Cohn (1990). "Child-mother attachment of six-year-olds and social competence at school," *Child Development*, 61(1), https://doi.org/10.1111/j.1467–8624.1990.tb02768.x.

Samuel Cohn (2021). "Coercive capitalism: introduction to the special issue," *Sociology of Development*, 7(2), 117–126, https://doi.org/10.1525/sod.2021.7.2.117.

Christopher Coker (2019). *The Rise of the Civilizational State*, Cambridge: Polity Press.
Lizzie Coles-Kemp, Debi Ashenden & Kieron O'Hara (2018). "Why should I? Cybersecurity, the security of the state and the insecurity of the citizen," *Politics and Governance*, 6(2), 41–48, https://doi.org/10.17645/pag.v6i2.1333.
Andrew Collins (2003). *Where Did It All Go Right? Growing Up Normal in the 70s*, London: Ebury Press.
Patricia Hill Collins (2017). "Intersectionality and epistemic injustice," in Ian James Kidd, José Medina & Gaile Pohlhaus Jr. (eds.), *The Routledge Handbook of Epistemic Injustice*, Abingdon: Routledge, 115–124.
Patrick Collison & Tyler Cowen (2019). "We need a new science of progress," *The Atlantic*, 30th July, 2019, https://www.theatlantic.com/science/archive/2019/07/we-need-new-science-progress/594946/.
Kate Connolly (2024). "'This is one fabulous party': Germany fans slowly catching Euro 2024 fever," *The Guardian*, 29th June, 2024, https://www.theguardian.com/football/article/2024/jun/29/germany-denmark-football-euro-2024.
Joseph Conrad (1912). *A Personal Record: Some Reminiscences*, New York: Harper & Brothers.
Gabriele Cosentino (2020). *Social Media and the Post-Truth World Order: The Global Dynamics of Disinformation*, Cham: Palgrave Pivot.
Emily Cousens (2020). "I teach at Oxford, but I don't want it to win the coronavirus vaccine race," *Huffpost*, 23rd Apr, 2020, https://www.huffingtonpost.co.uk/entry/coronavirus-vaccine_uk_5ea067f2c5b6b2e5b83ba372.
Tyler Cowen (2013). *Average is Over: Powering America Beyond the Age of the Great Stagnation*, New York: Dutton.
Joseph Cox (2024). *Dark Wire: The Incredible True Story of the Largest Sting Operation Ever*, New York: PublicAffairs.
Philip J. Cozzolino (2011). "Trust, cooperation, and equality: a psychological analysis of the formation of social capital," *British Journal of Social Psychology*, 50(2), 302–320, https://doi.org/10.1348/014466610X519610.
Kimberlé Crenshaw (2017). *On Intersectionality: Essential Writings*, New York: New Press.
Bernard Crick (1964). *In Defence of Politics*, revised edition, Harmondsworth: Pelican.
Kyle Crowder (2000). "The racial context of white mobility: an individual-level assessment of the white flight hypothesis," *Social Science Research*, 29(2), 223–257, https://doi.org/10.1006/ssre.1999.0668.
Steven Crowell (2022). "Commitment: what is self-binding, and how is it possible?" in Patricio A. Fernández, Alejandro Néstor García Martínez & José M. Torralba (eds.), *Ways of Being Bound: Perspectives from Post-Kantian Philosophy and Relational Sociology*, Cham: Springer, 29–46.
Raju J. Das (2023). "Why is identity politics not conducive to achieving sustained social justice?" *Dialectical Anthropology*, 47(1), 19–31, https://doi.org/10.1007/s10624–022–09673–6.
Sinclair Davidson, Primavera De Filippi & Jason Potts (2018). "Blockchains and the economic institutions of capitalism," *Journal of Institutional Economics*, 14 (4), 639–658, https://doi.org/10.1017/S1744137417000200.
Primavera De Filippi & Andrea Leiter (2021). "Blockchain in outer space," *American Journal of International Law Unbound*, 115, 413–418, https://doi.org/10.1017/aju.2021.63.

Primavera De Filippi, Morshed Mannan & Wessel Reijers (2020). "Blockchain as a confidence machine: the problem of trust and challenges of governance," *Technology in Society*, 62, 101284, https://doi.org/10.1016/j.techsoc.2020.101284.

Eduardo F.J. De Mulder (2019). 'Water', in Eduardo F.J. De Mulder, Ben C. De Pater & Joos C. Droogleever Fortuijn (eds.), *The Netherlands and the Dutch: A Physical and Human Geography*, Cham: Springer, 7–34.

Ithiel de Sola Pool (1963). "The mass media and politics in the modernization process," in Lucien W. Pye (ed.), *Communications and Political Development*, Princeton: Princeton University Press, 134–253.

Alexis de Tocqueville (2003). *Democracy in America*, London: Penguin.

Patrick J. Deneen (2019). *Why Liberalism Failed*, paperback edition, New Haven: Yale University Press.

Patrick J. Deneen (2023). *Regime Change: Towards a Postliberal Future*, New York: Penguin Random House.

Daniel C. Dennett (1987). *The Intentional Stance*, Cambridge MA: Bradford Books.

Karl Deutsch (1963). *The Nerves of Government: Models of Political Communication*, New York: Free Press.

Daniel Devine, Jennifer Gaskell, Will Jennings & Gerry Stoker (2021). "Trust and the coronavirus pandemic: what are the consequences of and for trust? An early review of the literature," *Political Studies Review*, 19(2), 274–285, https://doi.org/10.1177/1478929920948684.

Thierry Devos, Dario Spini & Shalom H. Schwartz (2002). "Conflicts among human values and trust in institutions," *British Journal of Social Psychology*, 41(4), 481–494, https://doi.org/10.1348/014466602321149849.

Robin DiAngelo (2018). *White Fragility: Why It's So Hard for White People to Talk About Racism*, Boston: Beacon Press.

Frank Dikötter (2013). *The Tragedy of Liberation: A History of the Chinese Revolution 1945–1957*, London: Bloomsbury.

Peter Thisted Dinesen, Merlin Schaeffer & Kim Mannemar Sønderskov (2020). "Ethnic diversity and social trust: a narrative and meta-analytical review," *Annual Review of Political Science*, 23, 441–465, https://doi.org/10.1146/annurev-polisci-052918–020708.

John Dunn (1984). "The concept of 'trust' in the politics of John Locke," in Richard Rorty, J.B. Schneewind & Quentin Skinner (eds.), *Philosophy in History*, Cambridge: Cambridge University Press, 279–301.

Ian Dunt (2023). *How Westminster Works … and Why It Doesn't*, London: Weidenfeld & Nicolson.

Economist (2015). '"The trust machine," *The Economist*, 31st Oct, 2015, https://www.economist.com/leaders/2015/10/31/the-trust-machine.

Economist (2021). "The most powerful people in crypto," *The Economist*, 18th Dec, 2021, https://www.economist.com/christmas-specials/2021/12/18/the-most-powerful-people-in-crypto.

Economist (2023). "Reckoning with slavery remains an elite project in Britain," *The Economist*, 17th Aug, 2023, https://www.economist.com/britain/2023/08/17/reckoning-with-slavery-remains-an-elite-project-in-britain.

Tom Egerton (2024). "External shocks," in Anthony Seldon & Tom Egerton (eds.), *The Conservative Effect 2010–2024: 14 Wasted Years?* Cambridge: Cambridge University Press, 37–70.

Mark Elliot, Kieron O'Hara, Charles Raab, Christine M. O'Keeffe, Elaine Mackey, Chris Dibben, Heather Gowans, Kingsley Purdam & Karen Mc Cullagh (2018).

"Functional anonymisation: personal data and the data environment," *Computer Law and Security Review*, 34(2), 204–221, https://doi.org/10.1016/j.clsr.2018.02.001.
Jon Elster (1998). 'Introduction', in Jon Elster (ed.), *Deliberative Democracy*, Cambridge: Cambridge University Press, 1–18.
Paul Embery (2021). *Despised: Why the Modern Left Loathes the Working Class*, Cambridge: Polity Press.
Erik H. Erikson (1963). *Childhood and Society*, 2nd edition, New York: Norton.
Lina Eriksson & Alan Hájek (2007). "What are degrees of belief?" *Studia Logica*, 86(2), 183–213, https://doi.org/10.1007/s11225–007–9059–4.
Gokhan Ertug, Julia Brennecke, Balázs Kovács & Tengjian Zou (2022). "What does homophily do? A review of the consequences of homophily," *Academy of Management Annals*, 16(1), 38–69, https://doi.org/10.5465/annals.2020.0230.
María Cristina Escudero (2024). "Institutional resistance: the case of the Chilean Convention 2021–22," *Global Constitutionalism*, 13(1), 192–199, https://doi.org/10.1017/S2045381723000291.
Virginia Eubanks (2019). *Automating Inequality: How High-Tech Tools Profile, Police, and Punish the Poor*, paperback edition, New York: Picador.
José Fajardo & Manuela Dantas (2018). "Understanding the impact of severe hyperinflation experience on current household investment behavior," *Journal of Behavioral and Experimental Finance*, 17, 60–67, https://doi.org/10.1016/j.jbef.2017.12.008.
John Fender (2020). *Austerity: When is it a Mistake and When is it Necessary?* Newcastle upon Tyne: Agenda.
Abby L. Ferber (2007). "Whiteness studies and the erasure of gender," *Sociology Compass*, 1(1), 265–282, https://doi.org/10.1111/j.1751–9020.2007.00014.x.
Niall Ferguson (2017). *The Square and the Tower: Networks, Hierarchies and the Struggle for Global Power*, London: Allen Lane.
Carmel Flaskas & Catherine Humphreys (1993). "Theorizing about power: intersecting the ideas of Foucault with the 'problem' of power in family therapy," *Family Process*, 32(1), 35–47, https://doi.org/10.1111/j.1545–5300.1993.00035.x.
Jennifer Fleetwood & John Lea (2022). "Defunding the police in the UK: critical questions and practical suggestions," *Howard Journal of Crime and Justice*, 61(2), 167–184, https://doi.org/10.1111/hojo.12468.
Elaine Forde (2017). "The ethics of energy provisioning: living off-grid in rural Wales," *Energy Research and Social Science*, 30, 82–93, https://doi.org/10.1016/j.erss.2017.06.018.
E.M. Forster (1972). "What I believe," in *Two Cheers for Democracy*, London: Edward Arnold, 65–73.
Diego Fossati, Burhanuddin Muhtadi & Eve Warburton (2022). "Why democrats abandon democracy: evidence from four survey experiments," *Party Politics*, 28(3), 554–566, https://doi.org/10.1177/1354068821992488.
Michel Foucault (1980). "Two lectures," in *Power/Knowledge: Selected Interviews and Other Writings 1972–1977*, Brighton: Harvester Press, 78–108.
Joe Foweraker (2016). "Democracy and its discontents in Latin America," in Joe Foweraker & Dolores Trevizo (eds.), *Democracy and Its Discontents in Latin America*, Boulder: Lynne Rienner Publishers, 1–14.
Free Exchange (2023). "Why the state should not promote marriage," *The Economist*, 28th Sept, 2023, https://www.economist.com/finance-and-economics/2023/09/28/why-the-state-should-not-promote-marriage.
Paolo Freire (2017). *Pedagogy of the Oppressed*, London: Penguin Random House.

Charles Fried (2015). *Contract As Promise: A Theory of Contractual Obligation*, 2nd edition, Oxford: Oxford University Press.
Jesse M. Fried (2021). "Will Nasdaq's diversity rules harm investors?" *SSRN*, https://dx.doi.org/10.2139/ssrn.3812642.
Thomas Fuchs (2010). "The psychopathology of hyperreflexivity," *Journal of Speculative Philosophy*, 24(3), 239–255.
Francis Fukuyama (1992). *The End of History and the Last Man*, London: Penguin.
Francis Fukuyama (1995). *Trust: The Social Virtues and the Creation of Prosperity*, New York: Free Press.
Manuel Funke, Moritz Schularick & Christoph Trebesch (2023). "Populist leaders and the economy," *American Economic Review*, 113(12), 3249–3288, https://doi.org/10.1257/aer.20202045.
Diego Gambetta (1988). "Can we trust trust?" in Diego Gambetta (ed.), *Trust: Making and Breaking Cooperative Relations*, Oxford: Blackwell, 213–237.
Diego Gambetta & Heather Hamill (2005). *Streetwise: How Taxi Drivers Establish Their Customers' Trustworthiness*, New York: Russell Sage Foundation.
William A. Gamson & Micah L. Sifry (2013). "The #Occupy Movement: an introduction," *Sociological Quarterly*, 54(2), 159–163, https://doi.org/10.1111/tsq.12026.
Steve Garner (2007). *Whiteness: An Introduction*, Abingdon: Routledge.
Mark Garnett (2004). *The Snake That Swallowed its Tail: Some Contradictions in Modern Liberalism*, Exeter: Imprint Academic.
Sophie Gaston (2018). *Citizens' Voices: Insights from Focus Groups Conducted in England for the Project*, At Home in One's Past project, London: Demos, https://www.demos.co.uk/project/citizens-voices/.
David Gauthier (1986). *Morals By Agreement*, Oxford: Oxford University Press.
Marianne George (2012). "Polynesian navigation and *Te Lapa* — 'the flashing'," *Time and Mind*, 5(2), 135–173, https://doi.org/10.2752/175169712X13294910382900.
Steven Anthony Gerencser (2000). *The Skeptic's Oakeshott*, New York: St Martin's Press.
Samuel J. Gershman, Eric J. Horwitz & Joshua B. Tenenbaum (2015). "Computational rationality: a converging paradigm for intelligence in brains, minds and machines," *Science*, 349(6245), 273–278, https://doi.org/10.1126/science.aac6076.
Marilena Geugjes (2021). *Collective Identity and Integration Policy in Denmark and Sweden*, Wiesbaden: Springer.
Bruce Cannon Gibney (2017). *A Generation of Sociopaths: How the Baby Boomers Betrayed America*, New York: Hachette.
Anthony Giddens (1984). *The Constitution of Society: Outline of the Theory of Structuration*, Cambridge: Polity Press.
Anthony Giddens (1990). *The Consequences of Modernity*, Cambridge: Polity Press.
Nuno Gil & Yongcheng Fu (2022). "Megaproject performance, value creation, and value distribution: an organizational governance perspective," *Academy of Management: Discoveries*, 8(2), 224–251, https://doi.org/10.5465/amd.2020.0029.
Margaret Gilbert (2014a). "Acting together," in *Joint Commitment: How We Make the Social World*, New York: Oxford University Press, 23–36.
Margaret Gilbert (2014b). "Who's to blame? Collective moral responsibility and its implications for group members," in *Joint Commitment: How We Make the Social World*, New York: Oxford University Press, 58–80.
George Gilder (2018). *Life After Google: The Fall of Big Data and the Rise of the Blockchain Economy*, Washington DC: Regnery Gateway.

John Gilliom (2001). *Overseers of the Poor: Surveillance, Resistance, and the Limits of Privacy*, Chicago: University of Chicago Press.

Catherine Gimbrone, Lisa M. Bates, Seth J. Prins & Katherine M. Keyes (2022). "The politics of depression: diverging trends in internalizing symptoms among US adolescents by political beliefs," *SSM – Mental Health*, 2, 100043, https://doi.org/10.1016/j.ssmmh.2021.100043.

Debbie Ging (2019). "Alphas, betas, and incels: theorizing the masculinities of the manosphere," *Men and Masculinities*, 22(4), 638–657, https://doi.org/10.1177/1097184X17706401.

René Girard (2023). 'Conflict', in *All Desire is a Desire for Being*, Dublin: Penguin, 3–4.

Erving Goffman (1959). *The Presentation of Self in Everyday Life*, New York: Doubleday.

Ian Goldin, Geoffrey Cameron & Meera Balarajan (2011). *Exceptional People: How Migration Shaped Our World and Will Define Our Future*, Princeton: Princeton University Press.

Agnieszka Golec de Zavala, Aleksandra Cichocka, Roy Eidelson & Nuwan Jayawickreme (2009). "Collective narcissism and its social consequences," *Journal of Personality and Social Psychology*, 97(6), 1074–1096, https://doi.org/10.1037/a0016904.

Agnieszka Golec de Zavala & Dorottya Lantos (2020). "Collective narcissism and its social consequences: the bad and the ugly," *Current Directions in Psychological Science*, 29(3), 273–278, https://doi.org/10.1177/0963721420917703.

David Goodhart (2017). *The Road to Somewhere: The Populist Revolt and the Future of Politics*, London: Hurst and Company.

Robert E. Goodin (1992). *Green Political Theory*, Cambridge: Polity Press.

Matthew Goodwin & Caitlin Milazzo (2017). "Taking back control? Investigating the role of immigration in the 2016 vote for Brexit," *British Journal of Politics and International Relations*, 19(3), 450–464, https://doi.org/10.1177/1369148117710799.

Robin Goodwin, Yaira Hamama-Raz, Elazar Leshem & Menachem Ben-Ezra (2023). "National resilience in Ukraine following the 2022 Russian invasion," *International Journal of Disaster Risk Reduction*, 85, 103487, https://doi.org/10.1016/j.ijdrr.2022.103487.

Richard Gott (2011). *Hugo Chávez and the Bolivarian Revolution*, London: Verso.

Trudy Govier & Wilhelm Verwoerd (2002). "Trust and the problem of national reconciliation," *Philosophy of the Social Sciences*, 32(2), 178–205, https://doi.org/10.1177/004931032002003.

David Graeber (2009). *Direct Action: An Ethnography*, Oakland: AK Press.

Matthew H. Graham & Milan W. Svolik (2020). "Democracy in America? Partisanship, polarization, and the robustness of support for democracy in the United States," *American Political Science Review*, 114(2), 392–409, https://doi.org/10.1017/S0003055420000052.

John Gray (1993a). "Oakeshott as a liberal," in *Post-Liberalism: Studies in Political Thought*, Abingdon: Routledge, 40–46.

John Gray (1993b). "What is dead and what is living in liberalism," in *Post-Liberalism: Studies in Political Thought*, Abingdon: Routledge, 283–328.

John Gray (1995). "The undoing of conservatism," in *Enlightenment's Wake: Politics and Culture at the Close of the Modern Age*, London: Routledge, 87–119.

John Gray (2000). *Two Faces of Liberalism*, Cambridge: Polity Press.

Robert Greally (2024). "Ideologies of political constitutionalism," *Oxford Journal of Legal Studies*, https://doi.org/10.1093/ojls/gqae028.
Harriet Green (2016). "Introducing the DAO: the organisation that will kill corporations," *City A.M.*, 3rd May, 2016, http://www.cityam.com/240198/introducing-the-dao-the-organisation-that-will-kill-corporations.
Andy Greenberg & Matt Burgess (2024). "The mystery of 'Jia Tan,' the XZ backdoor mastermind," *Wired*, 3rd Apr, 2024, https://www.wired.com/story/jia-tan-xz-backdoor/.
H.P. Grice (1989). "Logic and conversation," in *Studies in the Way of Words*, Cambridge MA: Harvard University Press, 22–40.
Jürgen Habermas (2012). *The Crisis of the European Union: A Response*, Cambridge: Polity Press.
Jürgen Habermas (2015). *The Lure of Technocracy*, Cambridge: Polity Press.
Nermin Hajdarbegovic (2014). "Bitcoin miners ditch ghash.io pool over fears of 51% attack," *CoinDesk*, 9th Jan, 2014, https://www.coindesk.com/markets/2014/01/09/bitcoin-miners-ditch-ghashio-pool-over-fears-of-51-attack/.
Maurice Halbwachs (1992). *On Collective Memory*, Chicago: University of Chicago Press.
Jonathan Hall & Katharina Werner (2022). "Trauma and trust: how war exposure shapes social and institutional trust among refugees," *Frontiers in Psychology*, 13, https://doi.org/10.3389/fpsyg.2022.786838.
Mary Katharine Ham & Guy Benson (2015). *End of Discussion: How the Left's Outrage Industry Shuts Down Debate, Manipulates Voters, and Makes America Less Free (and Fun)*, New York: Crown Forum.
Hansard Society (2017). *Audit of Political Engagement 14: The 2017 Report*, London: Hansard Society, http://doc.ukdataservice.ac.uk/doc/8183/mrdoc/pdf/8183_audit_of_political_engagement_14_2017.pdf.
Jaron Harambam & Stef Aupers (2017). "'I am not a conspiracy theorist': relational identifications in the Dutch conspiracy milieu," *Cultural Sociology*, 11(1), 113–129, https://doi.org/10.1177/1749975516661959.
Russell Hardin (1996). "Trustworthiness," *Ethics*, 107(1), 26–42, https://doi.org/10.1086/233695.
Russell Hardin (1999). "Do we want trust in government?" in Mark E. Warren (ed.), *Democracy and Trust*, Cambridge: Cambridge University Press, 22–41.
Russell Hardin (2006). *Trust*, Cambridge: Polity Press.
Gillian Harkins (2012). "Foucault, the family and the cold monster of neoliberalism," in Robbie Duschinsky & Leon Antonio Rocha (eds.), *Foucault, the Family and Politics*, London: Palgrave Macmillan, 82–117.
Vivien Hart (1978). *Distrust and Democracy: Political Distrust in Britain and America*, Cambridge: Cambridge University Press.
John Haugeland (1998). "Truth and rule-following," in *Having Thought: Essays in the Metaphysics of Mind*, Cambridge MA: Harvard University Press, 305–361.
Katherine Hawley (2014). "Trust, distrust and commitment," *Noûs*, 48(1), 1–20.
Katherine Hawley (2017). "Trust, distrust, and epistemic injustice," in Ian James Kidd, José Medina & Gaile Pohlhaus Jr. (eds.), *The Routledge Handbook of Epistemic Injustice*, Abingdon: Routledge, 69–78.
Katherine Hawley (2019). *How to be Trustworthy*, Oxford: Oxford University Press.
F.A. Hayek (1945). "The use of knowledge in society," *American Economic Review*, 35(4), 519–530.
F.A. Hayek (1960). *The Constitution of Liberty*, London: Routledge & Kegan Paul.

Bernd Hayo & Florian Neumaier (2016). "The debt brake in the eyes of the German population," *International Economics and Economic Policy*, 13(1), 139–159, https://doi.org/10.1007/s10368–015–0323–1.
Amanda Heitkamp & Thomas J. Mowen (2023). "Self-control and belief in conspiracy theories," *Social Currents*, 10(6), 503–511, https://doi.org/10.1177/23294965231193382.
Marc Hertogh (2018). *Nobody's Law: Legal Consciousness and Legal Alienation in Everyday Life*, London: Palgrave Macmillan.
Hermann Hesse (1972). "Zarathustra's return: a word to German youth," in *If the War Goes On: Reflections on War and Politics*, London: Jonathan Cape, 77–100.
Peter Hill (2014). "The Japanese Yakuza," in Letizia Paoli (ed.), *The Oxford Handbook of Organized Crime*, New York: Oxford University Press, 234–253.
Alison Hills (2023). "Trustworthiness, responsibility and virtue," *Philosophical Quarterly*, 73(3), 743–761, https://doi.org/10.1093/pq/pqad036.
Pekka Himanen (2001). *The Hacker Ethic and the Spirit of the Information Age*, London: Vintage.
Hillary Hiner (2022). "'We Didn't See It Coming': the rejection of the draft constitution reveals the powerful sway neoliberalism and nationalism continue to hold over Chilean politics and society," *NACLA Report on the Americas*, 54(4), 415–423, https://doi.org/10.1080/10714839.2022.2145121.
Afua Hirsch (2021). "We can mourn Prince Philip, but not the monarchy," *The Guardian*, 15th Apr, 2021, https://www.theguardian.com/commentisfree/2021/apr/15/we-can-mourn-prince-philip-but-not-the-monarchy.
Albert O. Hirschman (1982). *Shifting Involvements: Private Interest and Public Action*, Princeton: Princeton University Press.
Thomas Hobbes (1968). *Leviathan*, Harmondsworth: Penguin.
Pru Hobson-West (2003). "Understanding vaccination resistance: moving beyond risk," *Health, Risk and Society*, 5(3), 273–283, https://doi.org/10.1080/13698570310001606978.
Richard Hofstadter (2008a). "The paranoid style in American politics," in *The Paranoid Style in American Politics*, New York: Vintage Books, 3–40.
Richard Hofstadter (2008b). "The pseudo-conservative revolt – 1954," in *The Paranoid Style in American Politics*, New York: Vintage Books, 41–65.
Richard Hofstadter (2008c). "Goldwater and pseudo-conservative politics," in *The Paranoid Style in American Politics*, New York: Vintage Books, 93–141.
Martin Hollis (1998). *Trust Within Reason*, Cambridge: Cambridge University Press.
Jeremy Holmes (2014). *John Bowlby and Attachment Theory*, 2nd edition, Hove: Routledge.
Richard Holton (1994). "Deciding to trust, coming to believe," *Australasian Journal of Philosophy*, 72(1), 63–76, https://doi.org/10.1080/00048409412345881.
Ted Honderich (2005). *Conservatism: Burke, Nozick, Bush, Blair?* Revised edition, Pluto Press, London.
Christopher Hood (2011). *The Blame Game: Spin, Bureaucracy, and Self-Preservation in Government*, Princeton: Princeton University Press.
Christopher Hooks (2024). "Why did Greg Abbott pardon a racist murderer?" *TexasMonthly*, 17th May, 2024, https://www.texasmonthly.com/news-politics/greg-abbott-daniel-perry-pardon/.
Lars Hornuf, Theresa Kück & Armin Schwienbacher (2022). "Initial coin offerings, information disclosure, and fraud," *Small Business Economics*, 58(4), 1741–1759, https://doi.org/10.1007/s11187–021–00471-y.

Kimberly A. Houser (2019). "Can AI solve the diversity problem in the tech industry? Mitigating noise and bias in employment decision-making," *Stanford Technology Law Review*, 22, 290–354.

Philip N. Howard (2020). *Lie Machines: How to Save Democracy From Troll Armies, Deceitful Robots, Junk News Operations, and Political Operatives*, New Haven: Yale University Press.

Franz Huber & Christoph Schmidt-Petri (eds.) (2009). *Degrees of Belief*, Dordrecht: Springer.

Mike Hulme (2019). "Climate emergency politics is dangerous," *Issues in Science and Technology*, 36, (1), 23–25.

David Hume (1978). *A Treatise of Human Nature*, 2nd edition, Oxford: Oxford University Press.

David Hume (1987a). "Of the rise and progress of the arts and sciences," in *Essays: Moral, Political and Literary*, Indianapolis: Liberty Fund, 111–137.

David Hume (1987b). "Idea of a perfect commonwealth," in *Essays: Moral, Political and Literary*, Indianapolis: Liberty Fund, 512–529.

Mathew Humphrey (2006). "Democratic legitimacy, public justification and environmental direct action," *Political Studies*, 54(2), 310–327, https://doi.org/10.1111/j.1467–9248.2006.00602.x.

Samuel Huntington (1957). "Conservatism as an ideology," *American Political Science Review*, 51(2), 454–473, https://doi.org/10.2307/1952202.

Samuel Huntington (1996). *The Clash of Civilizations: And the Remaking of the World Order*, New York: Simon & Schuster.

Rosalind Hursthouse (1999). *On Virtue Ethics*, Oxford: Oxford University Press.

Luis-Daniel Ibáñez, Kieron O'Hara & Elena Simperl (2018). *On Blockchains and the General Data Protection Regulation*, EU Blockchain Observatory and Forum, https://www.eublockchainforum.eu/research-paper/blockchains-and-general-data-protection-regulation-gdpr.

Pablo Iglesias (2015). "Understanding Podemos," *New Left Review*, 93, 7–22.

Ghita Ionescu & Isabel de Madariaga (1972). *Opposition: Past and Present of a Political Institution*, Harmondsworth: Pelican.

Chris J. Jackson (2018). "Are survivalists malevolent?" *Personality and Individual Differences*, 129, 104–107, https://doi.org/10.1016/j.paid.2018.03.006.

Regina Jackson & Saira Rao (2022). *White Women: Everything You Already Know About Your Own Racism and How to Do Better*, New York: Penguin.

Marijn Janssen, Nripendra P. Rana, Emma L. Slade & Yogesh K. Dwivedi (2018). "Trustworthiness of digital government services: deriving a comprehensive theory through interpretive structural modelling," *Public Management Review*, 20(5), 647–671, https://doi.org/10.1080/14719037.2017.1305689.

Lars E.F. Johannessen (2019). "Negotiated discretion: redressing the neglect of negotiation in 'street-level bureaucracy'," *Symbolic Interaction*, 42(4), 513–538, https://doi.org/10.1002/symb.451.

Alan Johnson (2023). "'Progressives' and the Hamas pogrom: an A-Z guide," *Fathom*, 9th Oct, 2023, https://fathomjournal.org/progressives-and-the-hamas-pogrom-an-a-z-guide/.

James Johnson (2019). "The AI-cyber nexus: implications for military escalation, deterrence and strategic stability," *Journal of Cyber Policy*, 4(3), 442–460, https://doi.org/10.1080/23738871.2019.1701693.

Nevil Johnson (1980). "Constitutional reform: some dilemmas for a conservative philosophy," in Zig Layton-Henry (ed.), *Conservative Party Politics*, London: Macmillan, 126–155.

Paul Johnson, Carl Emmerson & Nick Ridpath (2024). "The economy," in Anthony Seldon & Tom Egerton (eds.), *The Conservative Effect 2010–2024: 14 Wasted Years?* Cambridge: Cambridge University Press, 71–101.

Karen Jones (1996). "Trust as an affective attitude," *Ethics*, 107(1), 4–25, https://doi.org/10.1086/233694.

Karen Jones (2012). "Trustworthiness," *Ethics*, 123(1), 61–85, https://doi.org/10.1086/667838.

Marin Jovanovic, Nikola Kostić, Ina M. Sebastian & Tomaz Sedej (2022). "Managing a blockchain-based platform ecosystem for industry-wide adoption: the case of TradeLens," *Technological Forecasting and Social Change*, 184, 121981, https://doi.org/10.1016/j.techfore.2022.121981.

John B. Judis & Ruy Teixeira (2002). *The Emerging Democratic Majority*, New York: Lisa Drew.

John B. Judis & Ruy Teixeira (2023). *Where Have All the Democrats Gone? The Soul of the Party in the Age of Extremes*, New York: Henry Holt.

Ari Juels, Ahmed Kosba & Elaine Shi (2016). "The ring of Gyges: investigating the future of criminal smart contracts," *Proceedings of the 2016 ACM SIGSAC Conference on Computer and Communications Security*, New York: ACM, 283–295, https://doi.org/10.1145/2976749.2978362.

Daniel Kahneman (2011). *Thinking, Fast and Slow*, New York: Farrar, Strauss & Giroux.

Keith Kahn-Harris (2018). *Denial: The Unspeakable Truth*, Mirefoot: Notting Hill Editions.

Elaine Kamarck (2024). "Abortion and the 2024 election: there is no easy way out for Republicans," *brookings.edu*, 17th Apr, 2024, https://www.brookings.edu/articles/abortion-and-the-2024-election-there-is-no-easy-way-out-for-republicans/.

Maria Kanal & Susan B. Rottman (2021). "Everyday agency: rethinking refugee women's agency in specific cultural contexts," *Frontiers in Psychology*, 12, https://doi.org/10.3389/fpsyg.2021.726729.

Melissa S. Kearney (2023). *The Two-Parent Privilege: How the Decline in Marriage Has Increased Inequality and Lowered Social Morality, and What We Can Do About It*, Chicago: University of Chicago Press.

John Kekes (1997). *Against Liberalism*, Ithaca: Cornell University Press.

John Kekes (1998). *A Case For Conservatism*, Ithaca: Cornell University Press.

Anthony Kelly (2020). "Recontextualising partisan outrage online: analysing the public negotiation of Trump support among American conservatives in 2016," *AI and Society*, https://doi.org/10.1007/s00146–020–01109–5.

Kevin Kenny (2013). *Diaspora: A Very Short Introduction*, New York: Oxford University Press.

Kathryn A. Kerns, Lisa Klepac & AmyKay Cole (1996). "Peer relationships and preadolescents' perceptions of security in the child-mother relationship," *Developmental Psychology*, 32(3), 457–466, https://psycnet.apa.org/doi/10.1037/0012–1649.32.3.457.

Anjum Khurshid (2020). "Applying blockchain technology to address the crisis of trust during the COVID-19 pandemic," *JMIR Medical Informatics*, 8(9), e20477, https://doi.org/10.2196/20477.

M. Tahir Kilavuz & Nathanael Gratias Sumaktoyo (2020). "Hopes and disappointments: regime change and support for democracy after the Arab Uprisings," *Democratization*, 27(5), 854–873, https://doi.org/10.1080/13510347.2020.1746766.

Haesoo Kim, HaeEun Kim, Juho Kim & Jeong-woo Jang (2022). "When does it become harassment?: an investigation of online criticism and calling out in Twitter," *Proceedings of the ACM on Human-Computer Interaction*, 6(2), article no.474, https://doi.org/10.1145/3555575.

Roger Kimball (2005). "The intellectual virtues: being critical," in Digby Anderson (ed.), *Decadence: The Passing of Personal Virtue and its Replacement by Political and Psychological Slogans*, London: Social Affairs Unit, 220–240.

Charlie Kimber (2023). "Rejoice as Palestinian resistance humiliates racist Israel," *Socialist Worker*, 9th Oct, 2023, https://socialistworker.co.uk/international/rejoice-as-palestinian-resistance-humiliates-racist-israel/.

Dori Kimel (2014). "Personal autonomy and change of mind in promise and in contract," in Gregory Klass, George Letsas & Prince Saprai (eds.), *Philosophical Foundations of Contract Law*, Oxford: Oxford University Press, 96–115.

Anthony King & Ivor Crewe (2013). *The Blunders of our Governments*, London: Oneworld.

Andrew Kirton (2020). "Matters of trust as matters of attachment security," *International Journal of Philosophical Studies*, 28(5), 583–602. https://doi.org/10.1080/09672559.2020.1802971.

Trevor I. Kiviat (2015). "Beyond Bitcoin: issues in regulating blockchain transactions," *Duke Law Journal*, 65(3), 569–608.

Markus Klems, Jacob Eberhardt, Stefan Tai, Steffen Härtlein, Simon Buchholz & Ahmed Tidjani (2017). "Trustless intermediation in blockchain-based decentralized service marketplaces," in Michael Maximilien, Antonio Vallecillo, Jianmin Wang & Marc Oriol (eds.), *Service-Oriented Computing: Proceedings of the 15th International Conference on Service-Oriented Computing (ICSOC)*, Cham: Springer, 731–739, https://doi.org/10.1007/978–3-319–69035–3_53.

Jonathan A. Knee (2021). *The Platform Delusion: Who Wins and Who Loses in the Age of Tech Titans*, New York: Portfolio.

Peter Knight (2000). *Conspiracy Culture: From Kennedy to the X-Files*, Abingdon: Routledge.

Kelli María Korducki (2020). "Accept that you're racist. Then, get to work dismantling racism," *medium.com*, 9th June, 2020, https://forge.medium.com/what-it-means-to-accept-that-youre-racist-fbeef3839e47.

Ivan Krastev & Stephen Holmes (2019). *The Light That Failed: A Reckoning*, New York: Penguin.

Thomas S. Kuhn (1970). *The Structure of Scientific Revolutions*, 2nd edition enlarged, Chicago: University of Chicago Press.

Jonatan Kurzwelly, Moira Pérez & Andrew D. Spiegel (2023). "Identity politics and social justice," *Dialectical Anthropology*, 47(1), 5–18, https://doi.org/10.1007/s10624–023–09686–9.

Ernesto Laclau (2005). *On Populist Reason*, London: Verso.

David Lammy (2011). "Mentors not Latin tutors," *The Guardian*, 29th Dec, 2011.

David Lammy (2019). "David Lammy on why there's nothing scary about a black man in a hoodie," *The Guardian*, 13th Feb, 2019, https://www.theguardian.com/world/2019/feb/13/david-lammy-on-why-theres-nothing-scary-about-a-black-man-in-a-hoodie.

Nick Land (2019). "Teleoplexy: notes on acceleration," in Robin Mackay & Armen Avanessian (eds.), *#Accelerate: The Accelerationist Reader*, 3rd edition, Falmouth: Urbanomic, 509–520.
Nancy K. Lankton, D. Harrison McKnight & John Tripp (2015). "Technology, humanness, and trust: rethinking trust in technology," *Journal of the Association for Information Systems*, 16(10), 880–918, https://doi.org/10.17705/1jais.00411.
Neal Lawson (2017). "A progressive alliance is an idea whose time has come," *The Guardian*, 23rd May, 2017, https://www.theguardian.com/commentisfree/2017/may/23/progressive-alliance-general-election-labour-future.
Neal Lawson (2021a). "Only a progressive alliance can defeat the Conservatives at the next election," *The Guardian*, 17th Feb, 2021, https://www.theguardian.com/commentisfree/2021/feb/17/progressive-alliance-defeat-conservatives-election.
Neal Lawson (2021b). "Here's how a progressive alliance would actually work," *The Guardian*, 24th Jun, 2021, https://www.theguardian.com/commentisfree/2021/jun/24/how-progressive-alliance-tory-britain.
Neal Lawson (2023). "There is a way to oust the Tories, and stop them ever rising again. Why won't we do it?" *The Guardian*, 9th May, 2023, https://www.theguardian.com/commentisfree/2023/may/09/way-to-oust-tories-local-elections-pr-proportional-representation.
Ryszard Legutko (2016). *The Demon in Democracy: Totalitarian Temptations in Free Societies*, New York: Encounter Books.
Hannes Leitgeb (2017). *The Stability of Belief: How Rational Belief Coheres With Probability*, Oxford: Oxford University Press.
Patti Tamara Lenard (2012). *Trust, Democracy, and Multicultural Challenges*, Pennsylvania: Pennsylvania State University Press.
Darryl R.J. Leroux & Adam Gaudry (2017). "Becoming Indigenous: the rise of Eastern Métis in Canada," *The Conversation*, 25th Oct, 2017, https://theconversation.com/becoming-indigenous-the-rise-of-eastern-metis-in-canada-80794.
Lawrence Lessig (1999). *Code: And Other Laws of Cyberspace*, New York: Basic Books.
Oliver Letwin (1999). *The Purpose of Politics*, London: The Social Market Foundation.
Margaret Levi (1998). "A state of trust," in Margaret Levi & Valerie Braithwaite (eds.), *Trust and Governance*, New York: Russell Sage Foundation.
Jacob S. Lewis & Sedef A. Topal (2021). *How does Exposure to Conflict Events Shape Social Trust? A Spatiotemporal Approach*, AfroBarometer working paper no.189, https://policycommons.net/artifacts/2177291/how-does-exposure-to-conflict-events-shape-social-trust-a-spatiotemporal-approach/2933168/.
Jacob Liedke & Jeffrey Gottfried (2022). "U.S. adults under 30 now trust information from social media almost as much as from national news outlets," *Pew Research Center*, 27th Oct, 2022, https://www.pewresearch.org/short-reads/2022/10/27/u-s-adults-under-30-now-trust-information-from-social-media-almost-as-much-as-from-national-news-outlets/.
Jonathan Liew (2023). "Ons Jabeur has shown how to live with defeat – like it's the end of the world," *The Guardian*, 18th July, 2023, https://www.theguardian.com/sport/2023/jul/18/ons-jabeur-live-with-defeat-wimbledon-tennis.
Orly Lobel (2022). *The Equality Machine: Harnessing Digital Technology for a Brighter, More Inclusive Future*, New York: PublicAffairs.
Glenn Loury, Winkfield Twyman Jr & Jennifer Richmond (2024). "The conundrum of colorblindness," *Substack*, 7th July, 2024, https://glennloury.substack.com/p/the-conundrum-of-colorblindness.

Jie Lu & Yun-han Chu (2021). "Trading democracy for governance," *Journal of Democracy*, 32(4), 115–130.

Niklas Luhmann (1988). "Familiarity, confidence, trust: problems and alternatives," in Diego Gambetta (ed.), *Trust: Making and Breaking Cooperative Relations*, Oxford: Blackwell, 94–107.

Niklas Luhmann (2017). *Trust and Power*, revised edition, Cambridge: Polity Press.

Steven Lukes (2019). "Trump's charisma," in Eric Klinenberg, Caitlin Zaloom & Sharon Marcus (eds.), *Antidemocracy in America: Truth, Power, and the Republic at Risk*, New York: Columbia University Press, 199–206.

Christopher Stephen Lutz (2012). *Reading Alasdair MacIntyre's* After Virtue, London: Continuum.

Naoíse Mac Sweeney (2023). *The West: A New History of an Old Idea*, London: W.H. Allen.

Alasdair MacIntyre (2007). *After Virtue*, 3rd edition, Notre Dame: University of Notre Dame Press.

Maersk (2022). *A.P. Moller-Maersk and IBM to Discontinue TradeLens, a Blockchain-Enabled Global Trade Platform*, A.P. Moller-Maersk press release, 29th Nov, 2022, https://www.maersk.com/news/articles/2022/11/29/maersk-and-ibm-to-discontinue-tradelens.

Alyson L. Mahar, Virginie Cobigo & Heather Stuart (2013). "Conceptualizing belonging," *Disability and Rehabilitation*, 35(12), 1026–1032, https://doi.org/10.3109/09638288.2012.717584.

Justin Malbon (2013). "Taking fake online consumer reviews seriously," *Journal of Consumer Policy*, 36(2), 139–157, https://doi.org/10.1007/s10603–012–9216–7.

Kenan Malik (2024). "Job ads aimed at the "benefits class" may be well-meant, but smack of contempt," *The Guardian*, 26th May, 2024, https://www.theguardian.com/commentisfree/article/2024/may/26/job-ads-aimed-at-the-benefits-class-may-be-well-meant-but-smack-of-contempt.

Sebastian Mallaby (2022). *The Power Law: Venture Capital and the Art of Disruption*, New York: Penguin.

Jane Mansbridge (1999). "Altruistic trust," in Mark E. Warren (ed.), *Democracy and Trust*, Cambridge: Cambridge University Press, 273–289.

Serena Marchesi, Davide Ghiglino, Francesca Ciardo, Jairo Perez-Osorio, Ebru Baykara & Agnieszka Wykowska (2019). "Do we adopt the intentional stance toward humanoid robots?" *Frontiers in Psychology*, 10, https://doi.org/10.3389/fpsyg.2019.00450.

Sam Margulies (2003). "The psychology of prenuptial agreements," *Journal of Psychology and Law*, 31(4), 415–432.

MaristPoll (2024). "A nation divided?" *MaristPoll*, 21st May, 2024, https://maristpoll.marist.edu/polls/a-nation-divided/.

Lorenzo Marsili & Niccolo Milanese (2018). *Citizens of Nowhere: How Europe Can Be Saved From Itself*, London: Zed.

Karl Marx & Friedrich Engels (2002). *The Communist Manifesto*, London: Penguin.

Abraham H. Maslow (1943). "A theory of human motivation," *Psychological Review*, 50(4), 370–396, https://doi.org/10.1037/h0054346.

Andrew Mason (2000). *Community, Solidarity and Belonging: Levels of Community and Their Normative Significance*, Cambridge: Cambridge University Press.

Diane K. Mauzy & R.S. Milne (2002). *Singapore Politics Under the People's Action Party*, London: Routledge.

Bruno Mazorra, Victor Adan & Vanesa Daza (2022). "Do not rug on me: leveraging machine learning techniques for automated scam detection," *Mathematics*, 10(6), article no.949, https://doi.org/10.3390/math10060949.
Mariana Mazzucato (2020). *Mission Economy: A Moonshot Guide to Changing Capitalism*, Dublin: Allen Lane.
Deirdre N. McCloskey (2006). *The Bourgeois Virtues: Ethics for an Age of Commerce*, Chicago: Chicago University Press.
Deirdre N. McCloskey (2010). *Bourgeois Dignity: Why Economics Can't Explain the Modern World*, Chicago: Chicago University Press.
Carole McGranahan (2017). "An anthropology of lying: Trump and the political sociality of moral outrage," *American Ethnologist*, 44(2), 243–248, https://doi.org/10.1111/amet.12475.
Graham Meikle (2023). *Deepfakes*, Cambridge: Polity Press.
Thomas Merton (2007). *New Seeds of Contemplation*, new edition, New York: New Directions.
Ines Michalowski (2011). "Required to assimilate? The content of citizenship tests in five countries," *Citizenship Studies*, 15(6–7), 749–768, https://doi.org/10.1080/13621025.2011.600116.
Christine Abdalla Mikhaeil & Richard L. Baskerville (2023). "Explaining online conspiracy theory radicalization: a second-order affordance for identity-driven escalation," *Information Systems Journal*, https://doi.org/10.1111/isj.12427.
John Stuart Mill (1991). "On liberty," in *On Liberty, and Other Essays*, Oxford: Oxford University Press, 1–128.
Michael F. Mills (2019). "Preparing for the unknown… unknowns: 'doomsday' prepping and disaster risk anxiety in the United States," *Journal of Risk Research*, 22(10), 1267–1279, https://doi.org/10.1080/13669877.2018.1466825.
R.S. Milne & Diane K. Mauzy (1999). *Malaysian Politics Under Mahathir*, London: Routledge.
Gabriel Salomon Mindel (2021). "The drums of Occupy," *Folklife Magazine*, 17th Sept, 2021, https://folklife.si.edu/magazine/occupy-movement-drum-circles.
Rachel Monaghan (1997). "Animal rights and violent protest," *Terrorism and Political Violence*, 9(4), 106–116, https://doi.org/10.1080/09546559708427432.
Rachel Monaghan (2013). "Not quite terrorism: animal rights extremism in the United Kingdom," *Studies in Conflict and Terrorism*, 36(11), 933–951, https://doi.org/10.1080/1057610X.2013.832117.
Montesquieu (1989). *The Spirit of the Laws*, Cambridge: Cambridge University Press.
T. Brian Mooney & Damini Roy (2021). "Politeness and *pietas* as annexed to the virtue of justice," in Chaoqun Xie (ed.), *The Philosophy of (Im)politeness*, Cham: Springer, 179–195.
Chantal Mouffe (2005). *On the Political*, Abingdon: Routledge.
Yascha Mounk (2022). *The Great Experiment: How to Make Diverse Democracies Work*, London: Bloomsbury.
Yascha Mounk (2023). *The Identity Trap: A Story of Ideas and Power in Our Time*, New York: Penguin.
Sisonke Msimang (2023). "I'm a white man who wants to signal allyship with people of colour. What should I do?" *The Guardian*, 12th May, 2023, https://www.theguardian.com/commentisfree/2023/may/13/im-a-white-man-who-wants-to-signal-allyship-with-people-of-colour-what-should-i-do.
Jan-Werner Müller (2017). *What is Populism?* New edition. London: Penguin.

Adnan Muminovic & Adnan Edfendic (2023). "The long-term effects of war exposure on generalized trust and risk attitudes: evidence from post-conflict Bosnia and Herzegovina," *Southeast European and Black Sea Studies*, 23(2), 299–316, https://doi.org/10.1080/14683857.2022.2121250.

Thomas Nagel (1974). "What is it like to be a bat?" *Philosophical Review*, 83(4), 435–450, https://doi.org/10.2307/2183914.

N. Chandrababu Naidu (2000). *Plain Speaking*, New Delhi: Viking.

Satoshi Nakamoto (2008). *Bitcoin: A Peer-to-Peer Electronic Cash System*, https://bitcoin.org/bitcoin.pdf.

Sofia Näsström (2021). *The Spirit of Democracy*, Oxford: Oxford University Press.

Anoop Nayak (2007). "Critical whiteness studies," *Sociology Compass*, 1(2), 737–755, https://doi.org/10.1111/j.1751-9020.2007.00045.x.

Susan Neiman (2023). *Left Is Not Woke*, Cambridge: Polity Press.

Friedrich Nietzsche (1967). *The Will to Power*, New York: Random House.

Friedrich Nietzsche (1973). *Beyond Good and Evil*, Harmondsworth: Penguin.

Friedrich Nietzsche (2007). *On the Genealogy of Morality*, revised student edition, Cambridge: Cambridge University Press.

William A. Niskanen Jr (1994). *Bureaucracy and Public Economics*, Cheltenham: Edward Elgar.

Helen Nissenbaum (2010). *Privacy in Context: Technology, Policy and the Integrity of Social Life*, Stanford: Stanford University Press.

Pippa Norris (2023). "Cancel culture: myth or reality?" *Political Studies*, 71(1), 145–174, https://doi.org/10.1177/00323217211037023.

Martha Nussbaum (2011). *Creating Capabilities: The Human Development Approach*, Cambridge MA: Harvard University Press.

J.C. Nyíri (1982). "Wittgenstein's later work in relation to conservatism," in Brian McGuinness (ed.), *Wittgenstein and his Times*, Oxford: Basil Blackwell, 44–68.

Kristóf (J.C.) Nyíri (2016). "Conservatism and common-sense realism," *The Monist*, 99(4), 441–456, https://doi.org/10.1093/monist/onw015.

Michael Oakeshott (1975). *On Human Conduct*, Oxford: Oxford University Press.

Michael Oakeshott (1991a). "Rationalism in politics," in *Rationalism in Politics and Other Essays*, Indianapolis: Liberty Fund, 5–42.

Michael Oakeshott (1991b). "Political education," in *Rationalism in Politics and Other Essays*, Indianapolis: Liberty Fund, 43–69.

Michael Oakeshott (1991c). "Introduction to *Leviathan*," in *Rationalism in Politics and Other Essays*, Indianapolis: Liberty Fund, 221–294.

Michael Oakeshott (1991d). "On being conservative," in *Rationalism in Politics and Other Essays*, Indianapolis: Liberty Fund, 407–437.

Michael Oakeshott (1996). *The Politics of Faith and the Politics of Scepticism*, New Haven: Yale University Press.

Sandra Obradović, Séamus A. Power & Jennifer Sheehy-Skeffington (2020). "Understanding the psychological appeal of populism," *Current Opinion in Psychology*, 35, 125–131, https://doi.org/10.1016/j.copsyc.2020.06.009.

Laurie T. O'Brien & Brenda Major (2009). "Group status and feelings of personal entitlement: the roles of social comparison and system-justifying beliefs," in John T. Jost, Aaron C. Kay & Hulda Thorisdottir (eds.), *Social and Psychological Bases of Ideology and System Justification*, New York: Oxford University Press, 427–443.

Kieron O'Hara (2004). *Trust: From Socrates to Spin*, Duxford: Icon.

Kieron O'Hara (2005). *After Blair: Conservatism Beyond Thatcher*, Thriplow: Icon.

Kieron O'Hara (2006). *The Referendum Roundabout*, Exeter: Imprint Academic.

Kieron O'Hara (2007). *Joseph Conrad Today*, Exeter: Imprint Academic.
Kieron O'Hara (2011). *Conservatism*, London: Reaktion.
Kieron O'Hara (2012). *A General Definition of Trust*, unpublished paper, https://kieronohara.com/article/a-general-definition-of-trust/.
Kieron O'Hara (2013). "The Conservative Dialectic of Margaret Thatcher's First Term," in Bradley W. Hart & Richard Carr (eds.), *The Foundations of the British Conservative Party*, London: Continuum, 39–61.
Kieron O'Hara (2015). "Authority printed upon emptiness," *IEEE Internet Computing*, 19(6), 72–76, https://doi.org/10.1109/MIC.2015.122.
Kieron O'Hara (2017a). *Why Does the Left Hate Trump? After All, In Some Ways He's One of Them*, ConservativeHome, 28th Jan, 2017, https://conservativehome.com/2017/01/28/kieron-ohara-why-does-the-left-hate-trump-after-all-in-some-ways-hes-one-of-them/.
Kieron O'Hara (2017b). "Smart contracts – dumb idea," *IEEE Internet Computing*, 21(2), 97–101, https://doi.org/10.1109/MIC.2017.48.
Kieron O'Hara (2020). "The contradictions of digital modernity," *AI and Society*, 35(1), 197–208, https://doi.org/10.1007/s00146–018–0843–7.
Kieron O'Hara (2021). "Burkean conservatism, legibility and populism," *Journal of Political Ideologies*, 26(1), 81–100, https://doi.org/10.1080/13569317.2020.1844371.
Kieron O'Hara (2022a). "Digital modernity," *Foundations and Trends in Web Science*, 9(1–2), 1–260, http://dx.doi.org/10.1561/1800000031.
Kieron O'Hara (2022b). "What conservatives value: reply to Blackburn," *Political Studies Review*, 20(3), 448–451, https://doi.org/10.1177/14789299211062039.
Kieron O'Hara (2023). *The Seven Veils of Privacy: How Our Debates About Privacy Conceal Its Nature*, Manchester: Manchester University Press.
Kieron O'Hara & Wendy Hall (2021). *Four Internets: Data, Geopolitics, and the Governance of Cyberspace*, New York: Oxford University Press.
Kieron O'Hara & David Stevens (2006). *inequality.com: Power, Poverty and the Digital Divide*, Oxford: Oneworld.
Jemima Olchawski (2024). "After 200 years, women can join the Garrick. Now for the task of making it share power, not hoard it," *The Guardian*, 8th May, 2024, https://www.theguardian.com/commentisfree/article/2024/may/08/women-join-garrick-club-vote-200-years-share-power.
Mancur Olson (1982). *The Rise and Decline of Nations: Economic Growth, Stagflation, and Social Rigidities*, New Haven: Yale University Press.
Onora O'Neill (2002). *A Question of Trust: The BBC Reith Lectures 2002*, Cambridge: Cambridge University Press.
Nicholas Orme (2021). *Going to Church in Medieval England*, New Haven: Yale University Press.
José Ortega y Gasset (1932). *The Revolt of the Masses*, New York: W.W. Norton.
Pierre Ostiguy & Kenneth M. Roberts (2016). "Putting Trump in comparative perspective: populism and the politicization of the sociocultural low," *Brown Journal of World Affairs*, 23(1), 25–50.
John O'Sullivan (2016). "Foreword," in Ryszard Legutko, *The Demon in Democracy: Totalitarian Temptations in Free Societies*, New York: Encounter Books, vii–x.
Adrian Pabst (2021). *Postliberal Politics*, Cambridge: Polity Press.
Camille Paglia (2017). "The nursery school campus: the corrupting of the humanities in the U.S.," in *Free Women Free Men: Sex, Gender, Feminism*, Edinburgh: Canongate, 78–84.

Michael Park, Erin Leahey & Russell J. Funk (2023). "Papers and patents are becoming less disruptive over time," *Nature*, 613, 138–144, https://doi.org/10.1038/s41586–022–05543-x.

ParliREACH (2019). *Stand In My Shoes: Race and Culture in Parliament*, London: ParliREACH, https://www.parliament.uk/globalassets/documents/foi/201041pr.pdf.

Volker Patent & Rosalind H. Searle (2019). "Qualitative meta-analysis of propensity to trust measurement," *Journal of Trust Research*, 9(2), 136–163, https://doi.org/10.1080/21515581.2019.1675074.

Harry Pearse (2020). "Deliberation, citizen science and Covid-19," *Political Quarterly*, 91(3), 571–577, https://doi.org/10.1111/1467–923X.12869.

J.E. Penner (2016). *The Law of Trusts*, 10th edition, Oxford: Oxford University Press.

Alex Pentland (2008). *Honest Signals: How They Shape Our World*, Cambridge MA: M.I.T. Press.

Philip Pettit (2004). "Trust, reliance and the Internet," *Analyse und Kritik*, 26(1), 108–121, https://doi.org/10.1515/auk-2004–0106.

Griffin M. Petty, Dustin Magilligan & Mandi Bates Bailey (2022). "You voted for *who*? Explaining support for Trump among racial and ethnic minorities," *New Political Science*, 44(2), 195–209, https://doi.org/10.1080/07393148.2022.2060004.

Guido Pezzarossi (2019). "Introduction: rethinking the archaeology of capitalism: coercion, violence, and the politics of accumulation," *Historical Archaeology*, 53(3–4), 453–467, https://doi.org/10.1007/s41636–019–00203-w.

James P. Pfiffner (2020). "The lies of Donald Trump: a taxonomy," in Charles M. Lamb & Jacob R. Neiheisel (eds.), *Presidential Leadership and the Trump Presidency: Executive Power and Democratic Government*, Cham: Palgrave Macmillan, 17–40.

Jenny Pickerill & John Krinsky (2012). "Why does Occupy matter?" *Social Movement Studies*, 11(3–4), 279–287, https://doi.org/10.1080/14742837.2012.708923.

Joseph M. Pierre (2024). "Conspiracies gone wild: a psychiatric perspective on conspiracy theory belief, mental illness, and the potential for lone actor ideological violence," *Terrorism and Political Violence*, https://doi.org/10.1080/09546553.2024.2329079.

Thomas Piketty (2021). ""Brahmin left versus merchant right: rising inequality and the changing structure of political conflict in France, the United States, and the United Kingdom, 1948–2020," in Amory Gethin, Clara Martínez-Toledano & Thomas Piketty (eds.), *Political Cleavages and Social Inequalities: A Study of Fifty Democracies, 1948–2020*, Cambridge MA: Harvard University Press, 85–135.

Fenne M. Pinkster & Willem R. Boterman (2017). "When the spell is broken: gentrification, urban tourism and privileged discontent in the Amsterdam canal district," *Cultural Geographies*, 24(3), 457–472, https://doi.org/10.1177/1474474017706176.

Hanna Fenichel Pitkin (1967). *The Concept of Representation*, Berkeley & Los Angeles: University of California Press.

Helen Pluckrose & James Lindsay (2020). *Cynical Theories: How Activist Scholarship Made Everything About Race, Gender and Identity – and Why This Harms Everybody*, London: Swift.

Peter Pomerantsev (2019). *This Is* Not *Propaganda: Adventures in the War Against Reality*, London: Faber & Faber.

Richard A. Posner (2014). *The Economic Analysis of Law*, 9th edition, New York: Wolters Kluwer Law & Business.

Nina Power (2022). *What Do Men Want? Masculinity and its Discontents*, London: Allen Lane.

Simon Price (2024). "Stop the celebrations – Oasis are the most damaging pop-cultural force in recent British history," *The Guardian*, 28th Aug, 2024, https://www.theguardian.com/commentisfree/article/2024/aug/28/stop-the-celebrations-oasis-are-the-most-damaging-pop-cultural-force-in-recent-british-history.

Robert D. Putnam (1993). *Making Democracy Work: Civic Traditions in Modern Italy*, Princeton: Princeton University Press.

Robert D. Putnam (2000). *Bowling Alone: The Collapse and Revival of American Community*, New York: Simon & Schuster.

Willard Van Orman Quine (1969). "Epistemology naturalized," in *Ontological Relativity and Other Essays*, New York: Columbia University Press, 69–90.

Josephine Quinn (2024). *How the World Made the West: A 4,000 Year History*, London: Bloomsbury.

John Rawls (1999). *A Theory of Justice*, revised edition, Cambridge, MA: Harvard University Press.

Eric S. Raymond (2001). *The Cathedral and the Bazaar: Musings on Linux and Open Source by an Accidental Revolutionary*, 2nd edition, Sebastopol CA: O'Reilly Media.

Jan Philipp Reemtsma (2012). *Trust and Violence*, Princeton: Princeton University Press.

José Renkens, Els Rommes & Maria van den Muijsenbergh (2022). "Refugees' agency: on resistance, resilience, and resources," *International Journal of Environmental Research and Public Health*, 19(2), https://www.mdpi.com/1660–4601/19/2/806.

Romina Green Rioja, Elisa Loncon & Alondra Carrillo Vidal (2022). "Transforming Chile from the ground up: two Constitutional Convention representatives reflect on the process of drafting the world's most progressive constitution," *NACLA Report on the Americas*, 54(4), 382–388, https://doi.org/10.1080/10714839.2022.2145114.

Kendrick B. Roberson (2024). "Insidious racism and institutional constraints: evidence from national and local case studies in the United States," *Humanities and Social Sciences Communications*, 11, article no.905, https://doi.org/10.1057/s41599–024–03419–1.

Alasdair Roberts (2012). "Why the Occupy movement failed," *Public Administration Review*, 72(5), 754–762.

Gareth Roberts (2024). *Gay Shame: The Rise of Gender Ideology and the New Homophobia*, London: Forum.

Hetty Roessingh, David Nordstokke & Mitchell Colp (2019). "Unlocking academic literacy in grade 4: the role of handwriting," *Reading and Writing Quarterly*, 35(2), 65–83, https://doi.org/10.1080/10573569.2018.1499160.

Pierre Rosanvallon (2008). *Counter-Democracy: Politics in an Age of Distrust*, Cambridge: Cambridge University Press.

Matthew Rose (2021). *A World After Liberalism: Five Thinkers Who Inspired the Radical Right*, New Haven: Yale University Press.

Morris Rosenberg (1956). "Misanthropy and political ideology," *American Sociological Review*, 21(6), 690–695, https://doi.org/10.2307/2088419.

Beate Rössler (2005). *The Value of Privacy*, Cambridge: Polity Press.

Mikael Rostila (2011). "The facets of social capital," *Journal for the Theory of Social Behaviour*, 41(3), 308–326, https://doi.org/10.1111/j.1468–5914.2010.00454.x.

Ken J. Rotenberg (2010). "The conceptualization of interpersonal trust: a basis, domain, and target framework," in Ken J. Rotenberg (ed.), *Interpersonal Trust During Childhood and Adolescence*, Cambridge: Cambridge University Press, 8–27.

Ken J. Rotenberg (2018), *The Psychology of Trust*, Abingdon: Routledge.

Ken J. Rotenberg, Michael J. Boulton & Claire L. Fox (2005). "Cross-sectional and longitudinal relations among children's trust beliefs, psychological maladjustment and social relationships: are very high as well as very low trusting children at risk?" *Journal of Abnormal Child Psychology*, 33(5), 595–610, https://doi.org/10.1007/s10802–005–6740–9.

Mike Rothschild (2021). *The Storm is Upon Us: How QAnon Became a Movement, Cult, and Conspiracy Theory of Everything*, London: Monoray.

Julian B. Rotter (1967). "A new scale for the measurement of interpersonal trust," *Journal of Personality*, 35(4), 651–665, https://psycnet.apa.org/doi/10.1111/j.1467–6494.1967.tb01454.x.

Olivier Roy (2004). *Globalized Islam: The Search For a New Ummah*, New York: Columbia University Press.

Giancarlo Ruffo, Alfonso Semeraro, Anastasia Giachanou & Paolo Rosso (2023). "Studying fake news spreading, polarisation dynamics, and manipulation by bots: a tale of networks and language," *Computer Science Review*, 47, 100531, https://doi.org/10.1016/j.cosrev.2022.100531.

Christopher F. Rufo (2023). *America's Cultural Revolution: How the Radical Left Conquered Everything*, New York: Broadside.

Stefan Rummens (2017). "Populism as a threat to liberal democracy," in Cristóbal Rovira Kaltwasser, Paul Taggart, Paulina Ochoa Espejo & Pierre Ostiguy (eds.), *The Oxford Handbook of Populism*, Oxford: Oxford University Press, 554–570.

Alexander Rüstow (2017). "General sociological causes of the economic disintegration and possibilities of reconstruction," in Thomas Biebricher & Frieder Vogelmann (eds.), *The Birth of Austerity: German Ordoliberalism and Contemporary Neoliberalism*, London: Rowman & Littlefield International, 151–162.

Lydia Saad (2023). "Historically low faith in U.S. institutions continues," *Gallup News*, 6th July, 2023, https://news.gallup.com/poll/508169/historically-low-faith-institutions-continues.aspx.

Stephen E. Sachs (2013). "Conflict resolution at a medieval English fair," in Albrecht Cordes, Serge Dauchy, Anika Auer & Andreas Karg (eds.), *Eine Grenze in Bewegung/Une Frontière Mouvante: Öffentliche und Private Justiz im Handels- und Seerecht/Justice Privée et Justice Publique en Matières Commerciales et Maritimes*, Munich: R. Oldenbourg Verlag, 19–38.

Joseph T. Salerno (2013). "Hyperinflation and the destruction of human personality," *Studia Humana*, 2(1), 15–27.

Joni Salminen, Chandrashekhar Kandpal, Ahmed Mohamed Kamel, Soon-gyo Jung & Bernard J. Jansen (2022). "Creating and detecting fake reviews of online products," *Journal of Retailing and Consumer Services*, 64, 102771, https://doi.org/10.1016/j.jretconser.2021.102771.

Michael J. Sandel (1996). *Democracy's Discontent: America in Search of a Public Philosophy*, Cambridge MA: Harvard University Press.

Michael J. Sandel (2021). *The Tyranny of Merit: What's Become of the Common Good?* Paperback edition, London: Penguin.

Jerry P. Sanson (2006). "'What he did and what he promised to do...': Huey Long and the horizons of Louisiana politics," *Louisiana History*, 47(3), 261–276.

Carl Schmitt (1988). *Political Theology*, Cambridge MA: M.I.T. Press.

Carl Schmitt (1996). *The Concept of the Political*, Chicago: University of Chicago Press.

Carl Schmitt (2008). *Constitutional Theory*, Durham NC: Duke University Press.

Ferdinand David Schoeman (1992). *Privacy and Social Freedom*, Cambridge: Cambridge University Press.
Florian Schui (2014). *Austerity: The Great Failure*, New Haven: Yale University Press.
Joseph Schumpeter (1950). *Capitalism, Socialism and Democracy*, 3rd edition, New York: Harper & Row.
Rachel Schurman & William A. Monro (2010). *Fighting for the Future of Food: Activists Versus Agribusiness in the Struggle Over Biotechnology*, Minneapolis: University of Minnesota Press.
Christine R. Schwartz (2013). "Trends and variation in assortative mating: causes and consequences," *Annual Review of Sociology*, 39, 451–470, https://doi.org/10.1146/annurev-soc-071312–145544.
James C. Scott (1998). *Seeing Like a State: How Certain Schemes to Improve the Human Condition Have Failed*, New Haven: Yale University Press.
James C. Scott (1999). "Geographies of trust, geographies of hierarchy," in Mark E. Warren (ed.), *Democracy and Trust*, Cambridge: Cambridge University Press, 273–289.
Roger Scruton (2015). *Fools, Frauds and Firebrands: Thinkers of the New Left*, London: Bloomsbury.
Roger Scruton (2016). *Philosophy: Principles and Problems*, London: Bloomsbury.
Roger Scruton (2017a). *Where We Are: The State of Britain Now*, London: Bloomsbury.
Roger Scruton (2017b). *Conservatism: An Invitation to the Great Tradition*, New York: All Points Books.
Roger Scruton (2017c). *On Human Nature*, Princeton University Press.
Roger Scruton (2022). "The art of taking offence," in Mark Dooley (ed.), *Against the Tide: The Best of Roger Scruton's Columns, Commentaries and Criticisms*, London: Bloomsbury Continuum, 85–87.
John R. Searle (2010). *Making the Social World: The Structure of Human Civilization*, Oxford: Oxford University Press.
Sophia Seawell (2021). "Meet the Black feminist politician shaking up Dutch politics," *OpenDemocracy*, 1st Jan, 2021, https://www.opendemocracy.net/en/5050/meet-first-black-woman-found-political-party-europe/.
Mark Sedgwick (2019). 'Introduction', in Mark Sedgwick (ed.), *Key Thinkers of the Radical Right: Behind the New Threat to Liberal Democracy*, New York: Oxford University Press, xiii–xxvi.
Anthony Seldon & Raymond Newell (2023). *Johnson at 10: The Inside Story*, London: Atlantic Books.
Adam B. Seligman (1997). *The Problem of Trust*, Princeton: Princeton University Press.
Amartya Sen (2010). *The Idea of Justice*, London: Penguin.
Nigel Shadbolt, Kieron O'Hara, David De Roure & Wendy Hall (2019). *The Theory and Practice of Social Machines*, Cham: Springer.
Ram Sewak Sharma (2020). *The Making of Aadhaar: World's Largest Identity Platform*, New Delhi: Rupa Publications.
Seana Valentine Shiffrin (2000). "Paternalism, unconscionability doctrine and accommodation," *Philosophy and Public Law*, 29(3), 205–250, https://doi.org/10.1111/j.1088–4963.2000.00205.x.
Seana Valentine Shiffrin (2007). "The divergence of contract and promise," *Harvard Law Review*, 120, 708–753.
Edward Shils (1997a). "Ideology and civility," in *The Virtue of Civility: Selected Essays on Liberalism, Tradition, and Civil Society*, Indianapolis: Liberty Fund, 25–62.

Edward Shils (1997b). "Tradition and liberty: antimony and interdependence," in *The Virtue of Civility: Selected Essays on Liberalism, Tradition, and Civil Society*, Indianapolis: Liberty Fund, 320–355.

Lionel Shriver (2024). "Kamala Harris is painfully out of her depth: the ineffectual VP isn't worthy of higher office," *UnHerd*, 22nd July, 2024, https://unherd.com/2024/07/kamala-harris-is-painfully-out-of-her-depth/.

Andrew Shryock (2008). "Thinking about hospitality, with Derrida, Kant, and the Balga Bedouin," *Anthropos*, 103(2), 405–421.

Larry Siedentop (2014). *Inventing the Individual: The Origins of Western Individualism*, London: Allen Lane.

Sergio Sismondo (2010). *An Introduction to Science and Technology Studies*, 2nd edition, Chichester: Wiley-Blackwell.

David Skelton (2019). *Little Platoons: How a Revived One Nation Can Empower England's Forgotten Towns and Redraw the Political Map*, London: Biteback.

Quentin Skinner (1995). "The paradoxes of political liberty," in Steven Darwall (ed.), *Equal Freedom: Selected Tanner Lectures on Human Values*, Ann Arbor: University of Michigan Press, 227–250.

Brian Skyrms (1996). *Evolution of the Social Contract*, Cambridge: Cambridge University Press.

Michael Slote (2001). *Morals From Motives*, Oxford: Oxford University Press.

Ralph Slovenko (1983). "The human/companion animal bond and the anthropomorphizing and naming of pets," *Medicine and Law*, 2, 277–283.

Martha J. Smith & Ronald V. Clarke (2000). "Crime and public transport," *Crime and Justice: A Review of Research*, 27, 169–233, https://doi.org/10.1086/652200.

Neil Smith (1996). *The New Urban Frontier: Gentrification and the Revanchist City*, London: Routledge.

Kelsey Smoot (2020). "White people say they want to be an ally to Black people. But are they ready for sacrifice?" *The Guardian*, 29th June, 2020, https://www.theguardian.com/commentisfree/2020/jun/29/white-people-ally-black-people-sacrifice.

Ernest Sosa (2007). *A Virtue Epistemology: Apt Belief and Reflective Knowledge, Volume I*, Oxford: Oxford University Press.

Andrew Sparrow (2014). "Nigel Farage: parts of Britain are 'like a foreign land'," *The Guardian*, 28th Feb, 2014, https://www.theguardian.com/politics/2014/feb/28/nigel-farage-ukip-immigration-speech.

Nick Spencer (2016). *The Evolution of the West: How Christianity Has Shaped Our Values*, London: Society for Promoting Christian Knowledge.

Gayatri Chakravorty Spivak (2006). "Subaltern studies: deconstructing historiography," in *In Other Worlds*, London: Routledge, 270–304.

Laurance Joseph Splitter (2022). *Identity, Reasonableness and Being* One Among Others*: Dialogue, Community, Education*, Singapore: Springer Nature Singapore Pte Ltd.

Timothy Stanley (2012). *The Crusader: The Life and Tumultuous Times of Pat Buchanan*, New York: Thomas Dunne Books.

Stefanie Stantcheva (2024). *Why Do We Dislike Inflation?* National Bureau of Economic Research working paper 32300, https://www.nber.org/papers/w32300.

Fred Steinmetz, Lennart Arne & Ingo Fiedler (2020). *Blockchain and the Digital Economy: the Socio-Economic Impact of Blockchain Technology*, Newcastle-Upon-Tyne: Agenda.

Kathleen Stock (2021). *Material Girls: Why Reality Matters for Feminism*, London: Fleet.

Luke Strongman (2008). "'When earth and sky almost meet': The conflict between traditional knowledge and modernity in Polynesian navigation," *Journal of World Anthropology: Occasional Papers*, 3(2), 48–110.

Alice Sullivan (2021). "Sex and the Office for National Statistics: a case study in policy capture," *Political Quarterly*, 92(4), 638–651, https://doi.org/10.1111/1467–923X.13029.

Cass R. Sunstein (2021). *Liars: Falsehoods and Free Speech in an Age of Deception*, New York: Oxford University Press.

Christine Swanton (2019). "The relation between virtue ethics and virtue epistemology," in Heather Battaly (ed.), *The Routledge Handbook of Virtue Epistemology*, New York: Routledge, 508–521.

Christine Swanton (2021). *Target Centred Virtue Ethics*, Oxford: Oxford University Press.

Rachel Sylvester (2024). "Health," in Anthony Seldon & Tom Egerton (eds.), *The Conservative Effect 2010–2024: 14 Wasted Years?* Cambridge: Cambridge University Press, 131–156.

Nick Szabo (1997). "Formalizing and securing relationships on public networks," *First Monday*, 2(9), https://doi.org/10.5210/fm.v2i9.548.

Mariarosaria Taddeo (2010). "Modelling trust in artificial agents, a first step toward the analysis of e-trust," *Minds and Machines*, 20(2), 243–257, https://doi.org/10.1007/s11023–010–9201–3.

Joshua Tate (2019). "Mencius Moldbug and neoreaction," in Mark Sedgwick (ed.), *Key Thinkers of the Radical Right: Behind the New Threat to Liberal Democracy*, New York: Oxford University Press, 187–203.

Wilson T.V. Tay (2017). "Dimensions of *Ketuanan Melayu* in the Malaysian constitutional framework," in Andrew Harding & Dian A.H. Shah (eds.), *Law and Society in Malaysia: Pluralism, Religion and Ethnicity*, London: Routledge, 44–71.

Charles Taylor (1985). "What is human agency?" in *Human Agency and Language: Philosophical Papers 1*, Cambridge: Cambridge University Press, 15–44.

Chloë Taylor (2012). "Foucault and familial power," *Hypatia*, 27(1), 201–218, https://doi.org/10.1111/j.1527–2001.2011.01171.x.

Audrey Thompson (2003). "Tiffany, friend of people of color: white investments in antiracism," *International Journal of Qualitative Studies in Education*, 16(1), 7–29, https://doi.org/10.1080/0951839032000033509.

Paul Thompson, Frederick Harry Pitts, Jo Ingold & Jon Cruddas (2022). "Class composition, Labour's strategy and the politics of work," *Political Quarterly*, 93(1), 142–149, https://doi.org/10.1111/1467–923X.13097.

Zhiyi Tian, Lei Cui, Jie Liang & Shui Yu (2022). "A comprehensive survey on poisoning attacks and countermeasures in machine learning," *ACM Computing Surveys*, 55(8), article no.166, https://doi.org/10.1145/3551636.

Michael Tichelar (2017). *The History of Opposition to Blood Sports in Twentieth Century England: Hunting at Bay*, Abingdon: Routledge.

Nicholas Timmins (2001). *The Five Giants: A Biography of the Welfare State*, revised paperback edition, London: HarperCollins.

Nick Timothy (2020). *Remaking One Nation: The Future of Conservatism*, Cambridge: Polity Press.

Anna Tims (2023). "Opodo and Austrian Airlines wrecked Ukrainian refugees' journey home," *The Guardian*, 8th Mar, 2023, https://www.theguardian.com/money/2023/mar/08/opodo-and-austrian-airlines-wrecked-ukrainian-refugees-journey-home.

Milind Tiwari, Adrian Gepp & Kuldeep Kumar (2020). "The future of raising finance – a new opportunity to commit fraud: a review of initial coin offering (ICOs) scams," *Crime, Law and Social Change*, 73(4), 417–441, https://doi.org/10.1007/s10611–019–09873–2.

Briana Toole (2022). "Demarginalizing standpoint epistemology," *Episteme*, 19(1), 47–65, https://doi.org/10.1017/epi.2020.8.

Richard C.S. Trahair (1999). *Utopias and Utopians: An Historical Dictionary*, Westport CT: Greenwood Press.

Laurence Trautman (2014). "Virtual currencies: Bitcoin & what now after Liberty Reserve, Silk Road, and Mt Gox?" *Richmond Journal of Law and Technology*, 20(4), article no.13.

TRM Labs (2023). *Global Crypto Policy Review & Outlook 2023/24*, https://www.trmlabs.com/post/trm-labs-finds-link-between-stronger-regulation-and-reduced-illicit-activity-in-new-report.

Nicholas Dana Troutman & Aron Laszka (2021). "PoolParty: efficient blockchain-agnostic decentralized mining pool," in *ICBCT '21: 2021 The 3rd International Conference on Blockchain Technology*, New York: ACM, 20–27, https://doi.org/10.1145/3460537.3460554.

Arianna Trozze, Josh Kamps, Eray Arda Akartuna, Florian J. Hetzel, Bennett Kleinberg, Toby Davies & Shane D. Johnson (2022). "Cryptocurrencies and future financial crime," *Crime Science*, 11, article no.1, https://doi.org/10.1186/s40163–021–00163–8.

Barbara W. Tuchman (2017). *A Distant Mirror: The Calamitous 14th Century*, paperback edition, London: Penguin.

Andrew Tucker (1999). *Why Trust Has No Part in Modern Politics*, London: Centre for Reform.

Aviezer Tucker (2008). "Pre-emptive democracy: oligarchic tendencies in deliberative democracy," *Political Studies*, 56(1), 127–147, https://doi.org/10.1111/j.1467–9248.2007.00676.x.

Peter Turchin (2023). *End Times: Elites, Counter-Elites and the Path of Political Disintegration*, New York: Penguin.

Stephen Turner (1994). *The Social Theory of Practices: Tradition, Tacit Knowledge and Presuppositions*, Cambridge: Polity Press.

Nadia Urbinati (2015). *The Tyranny of the Moderns*, New Haven: Yale University Press.

Nadia Urbinati & Mark E. Warren (2008). "The concept of representation in contemporary democratic theory," *Annual Review of Political Science*, 11, 387–412, https://doi.org/10.1146/annurev.polisci.11.053006.190533.

Eric M. Uslaner (1993). *The Decline of Comity in Congress*, Ann Arbor: University of Michigan Press.

Eric M. Uslaner (2002). *The Moral Foundations of Trust*, Cambridge: Cambridge University Press.

Tina van der Linden & Tina Shirazi (2023). "Markets in crypto-assets regulation: does it provide legal certainty and increase adoption of crypto-assets?" *Financial Innovation*, 9, article no.22, https://doi.org/10.1186/s40854–022–00432–8.

Jacquelien van Stekelenburg (2012). "The Occupy Movement: product of this time," *Development*, 55(2), 224–231, https://doi.org/10.1057/dev.2012.6.

Frans van Waarden (2012). "The governance of markets: on generating trust in transactions," in David Levi-Faur (ed.), *The Oxford Handbook of Governance*, Oxford: Oxford University Press, 355–371.

Yaron Velner, Jason Teutsch & Loi Luu (2017). "Smart contracts make Bitcoin mining pools vulnerable," in Michael Brenner, Kurt Rohloff, Joseph Bonneau, Andrew Miller, Peter Y.A. Ryan, Vanessa Teague, Andrea Bracciali, Massimiliano Sala, Federico Pintore & Markus Jakobsson (eds.), *Financial Cryptography and Data Security: FC 2017 International Workshops, WAHC, BITCOIN, VOTING, WTSC, and TA, Revised Selected Papers*, Cham: Springer, 298–316, https://doi.org/10.1007/978-3-319-70278-0_19.

Alan Wager (2021). "'Since this is a document of record': collecting the oral histories of the Brexit Parliament," *Political Quarterly*, 92(3), 552–558, https://doi.org/10.1111/1467-923X.13031.

George Walden (2020). *The New Elites: A Career in the Masses*, revised and updated edition, London: Gibson Square.

Jim Waldo (2019). "A hitchhiker's guide to the blockchain universe," *Communications of the ACM*, 62(3), 38–42, http://dx.doi.org/10.1145/3303868.

Karen Walker (1995). "'Always there for me': friendship patterns and expectations among middle- and working-class men and women," *Sociological Forum*, 10(2), 273–296, https://doi.org/10.1007/BF02095961.

Michael Walzer (1983). *Spheres of Justice: A Defense of Pluralism and Equality*, New York: Basic Books.

Mark E. Warren (1999). "Democratic theory and trust," in Mark E. Warren (ed.), *Democracy and Trust*, Cambridge: Cambridge University Press, 310–345.

Everett Waters & Kathleen E. Deane (1985). "Defining and assessing individual differences in attachment relationships: Q-methodology and the organization of behavior in infancy and early childhood," *Monographs of the Society for Research in Child Development*, 50(1–2), 41–65, https://doi.org/10.2307/3333826.

Simone Weil (1952). *The Need for Roots*, London: Routledge & Kegan Paul.

Stephanie Wescott, Steven Roberts & Xuenan Zhao (2024). "The problem of anti-feminist 'manfluencer' Andrew Tate in Australian schools: women teachers' experiences of resurgent male supremacy," *Gender and Education*, 36(2), 167–182, https://doi.org/10.1080/09540253.2023.2292622.

Laura Westra (1998). *Living in Integrity: A Global Ethic to Restore a Fragmented Earth*, Lanham MD: Rowman & Littlefield.

Kurt Weyland (2001). "Clarifying a contested concept: populism in the study of Latin American politics," *Comparative Politics*, 34(1), 1–22.

Geoff White (2024). *Rinsed: From Cartels to Crypto: How the Tech Industry Washes Money For the World's Deadliest Crooks*, Dublin: Penguin Business.

Theodore H. White (1962). *The Making of the President 1960*, London: Jonathan Cape.

David Willetts (2010). *The Pinch: How the Baby Boomers Took Their Children's Future – and Why they Should Give it Back*, London: Atlantic.

Alex Williams & Nick Srnicek (2019). "#Accelerate: manifesto for an accelerationist politics," in Robin Mackay & Armen Avanessian (eds.), *#Accelerate: The Accelerationist Reader*, 3rd edition, Falmouth: Urbanomic, 347–362.

Bernard Williams (1988). "Formal structures and social reality," in Diego Gambetta (ed.), *Trust: Making and Breaking Cooperative Relations*, Oxford: Blackwell, 3–13.

Thomas Chatterton Williams (2019). *Self Portrait in Black and White: Unlearning Race*, New York: W.W. Norton.

Timothy Williamson (2000). *Knowledge and its Limits*, Oxford: Oxford University Press.

Erika K. Wilson (2019). "The new white flight," *Duke Journal of Constitutional Law and Public Policy*, 14, 233–284.

Erik T. Withers (2017). "Whiteness and culture," *Sociology Compass*, 11(4), e12464, https://doi.org/10.1111/soc4.12464.
Ludwig Wittgenstein (1958). *Philosophical Investigations*, 2nd edition, Oxford: Basil Blackwell.
Ludwig Wittgenstein (1978). *Remarks on the Foundations of Mathematics*, 3rd edition, Oxford: Basil Blackwell.
Ludwig Wittgenstein (1979). *On Certainty*, reprint with corrections, Oxford: Basil Blackwell.
Ludwig Wittgenstein (1980). *Remarks on the Philosophy of Psychology*, Volume I, Oxford: Basil Blackwell.
Ludwig Wittgenstein (1981). *Zettel*, 2nd edition, Oxford: Basil Blackwell.
Ludwig Wittgenstein (1998). *Culture and Value*, revised edition, Malden MA: Blackwell.
Ludwig Wittgenstein (2009). "Philosophy of psychology – a fragment," in *Philosophical Investigations*, revised 4th edition, Chichester: Wiley-Blackwell, 182–243.
Rachael A. Woldoff (2011). *White Flight/Black Flight: The Dynamics of Racial Change in an American Neighborhood*, Ithaca: Cornell University Press.
Jonathan Wolff (2020). "Rhodes' statue is going. Now universities must ask: what about John Locke, Elizabeth I and others?" *The Guardian*, 7th July, 2020, https://www.theguardian.com/education/2020/jul/07/rhodes-statue-is-going-now-universities-must-ask-what-about-john-locke-elizabeth-i-and-others.
Freya A. Woods & Janet B. Ruscher (2021). "'Calling-out' vs. 'calling-in' prejudice: confrontation style affects inferred motive and expected outcomes," *British Journal of Social Psychology*, 60(1), 50–73, https://doi.org/10.1111/bjso.12405.
Moran Yarchi, Christian Baden & Neta Kligler-Vilenchik (2021). "Political polarization on the digital sphere: a cross-platform, over-time analysis of interactional, positional, and affective polarization on social media," *Political Communication*, 38(1–2), 98–139, https://doi.org/10.1080/10584609.2020.1785067.
Clifford Young, Katie Ziemer & Chris Jackson (2019). "Explaining Trump's popular support: validation of a nativism index," *Social Science Quarterly*, 100(2), 412–418, https://doi.org/10.1111/ssqu.12593.
Dannagal Goldthwaite Young (2020). *Irony and Outrage: The Polarized Landscape of Rage, Fear, and Laughter in the United States*, New York: Oxford University Press.
Bernardo Zacka (2017). *When the State Meets the Street: Public Service and Moral Agency*, Cambridge MA: Belknap Press.
Alex Zamalin (2021). *Against Civility: The Hidden Racism in Our Obsession With Civility*, Boston MA: Beacon.
Rainer Zitelmann (2019). *The Power of Capitalism: A Journey Through Recent History Across Five Continents*, London: LID Publishing.
Slavoj Žižek (2008). *Violence: Six Sideways Reflections*, London: Profile.
Shoshana Zuboff (2019). *The Age of Surveillance Capitalism: The Fight for a Human Future at the New Frontier of Power*, London: Profile.
Aizatul Anis binti Zuhari & Suyatno Ladiqi (2019). "Identity and *Ketuanan Melayu* in Malaysia: effects of globalization and challenges to the Ketuanan Melayu," in Fadhila Inas Pratiwi, Mohd Afandi Salleh, Mu'ammar Alif Zarkasi Sukma Raga, Iqbal Bagus Alfiansyah & Nanda Pradina Putri (eds.), *The Role of Identity In Politics and Policy Making*, Surabaya: Revka Prima Media, 71–82.

Index